T0193297

Praise for *Present Shock*

"Genius." —Tom Keen, BloombergTV

"In the interest of the perpetual now, I'll cut to the chase: you should read this book." —Chris Lites, *The Rumpus*

"In this refreshing antidote to promises of digital Utopia, Rushkoff articulates his own well-informed second thoughts. We should pay close attention—while we still can."

—George Dyson, author of *Turing's Cathedral and Darwin Among the Machines*

"Douglas Rushkoff is one of the rare critics of today's digitally colonized way of life who is both a user and maker of the media he critiques. *Present Shock* holds up new lenses and offers new narratives about what might be happening to us and why, compelling readers to look at the larger repercussions of today's technologically mediated social practices."

—Howard Rheingold, author of *Net Smart*

"A wide-ranging social and cultural critique, *Present Shock* artfully weaves through so many different kinds of materials as it makes its point: we are exhilarated, drugged, and consumed by the now."

—Sherry Turkle, professor of the Social Studies of Science, MIT; author of *Alone Together*

"With brilliant insight Rushkoff once again gets there early, making us confront the new world of 'presentism'—the shift in our focus from the future to the present, from the horizon-gazing to the experience of here and now."

—Marina Gorbis, executive director, Institute for the Future

"If you read one book next year to help you make sense of the present moment, let it be *Present Shock* by Douglas Rushkoff."

—Anthony Wing Kosner, Forbes.com

"A sobering wake-up call to collectively reexamine our relationship with time before we're blindsided by an unwelcome future." —*Booklist*

ABOUT THE AUTHOR

Douglas Rushkoff, PhD, is a world-renowned media theorist whose twelve books, including *Life Inc.* and *Program or Be Programmed,* have won prestigious awards and have been translated into thirty languages. He is a commentator on CNN and a contributor to *The Guardian* (London), Discover, and NPR. He also made the PBS documentaries *The Merchants of Cool*, *The Persuaders*, and *Digital Nation*. He lectures about technology and change at universities and conferences around the world. He lives in New York with his wife, Barbara, and daughter, Mamie.

Visit www.rushkoff.com.

PRESENT
SHOCK

WHEN
EVERYTHING
HAPPENS
NOW

DOUGLAS
RUSHKOFF

CURRENT

CURRENT

Published by the Penguin Group
Penguin Group (USA) LLC
375 Hudson Street
New York, New York 10014

USA | Canada | UK | Ireland | Australia | New Zealand | India | South Africa | China
penguin.com
A Penguin Random House Company

First published in the United States of America by Current, a member of Penguin Group (USA) Inc., 2013
This paperback edition published 2014

THE LIBRARY OF CONGRESS HAS CATALOGED THE HARDCOVER EDITION AS FOLLOWS:
Rushkoff, Douglas.
Present shock : when everything happens now / Douglas Rushkoff.
pages cm
Includes bibliographical references and index.
ISBN 978-1-59184-476-1 (hc.)
ISBN 978-1-61723-010-3 (pbk.)
1. Technology—Social aspects. 2. Technology—Philosophy. I. Title.
T14.5.R86 2013
303.48'3—dc23   2012039915

Set in Garamond 3 LT Std
Designed by Sabrina Bowers

146122990

For my daughter, Mamie, my present

CONTENTS

PREFACE

He is one of the most prescient hedge fund managers on Wall Street, but his trades always seem to happen after the fact. That's because as soon as he executes an order, it is observed and preempted by traders at bigger firms with faster computers. The spread changes, and his buy order goes through just a few fractions of a penny higher than it should have. He is trading in the past, longing for the software and geeks he needs to get into his competitors' present. And his clients can no longer conceive of investing in a company's future, anyway; they want to win on the trade itself, as it actually happens.

She's at a bar on Manhattan's Upper East Side, but she seems oblivious to the boys and the music. Instead of engaging with those around her, she's scrolling through text messages on her phone, from friends at other parties across town. She needs to know if the event she's at is the event to be at, or whether something better is happening at that very moment, somewhere else. Sure enough, a blip on the tiny screen catches her interest, and in seconds her posse is in a cab headed for the East Village. She arrives at a seemingly identical party and decides it's "the place to be," yet instead of enjoying it, she turns her phone around, activates the camera, and proceeds to take pictures of herself and her friends for the next hour—instantly uploading them for the world to see her in the moment.

He sees the signs all around him: the latest "natural" disaster on the evening news; the fluctuations in the prices at the gas pump; talk of a single world currency. Information overload might not have increased the rate at which disasters occur, but it has exponentially increased the rate at which they're witnessed. As a result, prophecy no longer feels like a description of the future but, rather, a guide to the present. The ideas of quantum physicists and the Mayans have been twisted to indicate that time itself will soon be coming to an end, anyway. The messianic age is no longer something to prepare for; it is a current event. What would Jesus do?

This is the new "now."

Our society has reoriented itself to the present moment. Everything is live, real time, and always-on. It's not a mere speeding up, however much our lifestyles and technologies have accelerated the rate at which we attempt to do things. It's more of a diminishment of anything that isn't happening right now—and the onslaught of everything that supposedly is.

It's why the world's leading search engine is evolving into a live, customized, and predictive flow of data branded "Google Now"; why email is giving way to texting, and why blogs are being superseded by Twitter feeds. It's why kids in school can no longer follow linear arguments; why narrative structure collapsed into reality TV; and why we can't engage in meaningful dialogue about last month's books and music, much less long-term global issues. It's why an economy once based on long-term investment and interest-bearing currency can no longer provide capital to those who plan to put it to work for future rewards. It's why so many long for a "singularity" or a 2012 apocalypse to end linear time altogether and throw us into a posthistoric eternal present—no matter the cost to human agency or civilization itself.

But it's also how we find out what's happening on the streets of Iran before CNN can assemble a camera crew. It's what enables an unsatisfied but upwardly mobile executive to quit his job and move with his family to Vermont to make kayaks—which he thought he'd get to do only once he retired. It's how millions of young people can choose to embody a new activism based in patient consensus instead of contentious debate. It's what enables companies like H&M or Zara to fabricate clothes in real time, based on the instantaneous data coming from scanned tags at checkout counters five thousand miles away. It's how a president can run for office and win by breaking from the seeming tyranny of the past and its false hope, and tell voters that "we are the ones we have been waiting for."

Well, the waiting is over. Here we are.

If the end of the twentieth century can be characterized by futurism, the twenty-first can be defined by presentism.

The looking forward so prevalent in the late 1990s was bound to end once the new millennium began. Like some others of that era, I predicted a new focus on the moment, on real experience, and on what things are actually worth right now. Then 9/11 magnified this sensibility, forcing America as a nation to contend with its own impermanence. People had babies in droves,[1] and even filed for divorces,[2] in what was at least an unconscious awareness that none of us lives forever and an accompanying reluctance to postpone things indefinitely. Add real-time technologies, from the iPhone to Twitter; a disposable consumer economy where 1-Click ordering is more important than the actual product being purchased; a multitasking brain actually incapable of storage or sustained argument; and an economy based on spending now what one may or may not earn in a lifetime, and you can't help but become temporally disoriented. It's akin to the onslaught of changing rules and circumstances that 1970s futurist Alvin Toffler dubbed "future shock."

Only, in our era it's more of a *present shock*. And while this phenomenon is clearly "of the moment," it's not quite as *in* the moment as we may have expected.

For while many of us were correct about the way all this presentism would affect investments and finance, even technology and media, we were utterly wrong about how living in the "now" would end up impacting us as people. Our focus on the present may have liberated us from the twentieth century's dangerously compelling ideological narratives. No one—well, hardly anyone—can still be convinced that brutal means are justified by mythological ends. And people are less likely to believe employers' and corporations' false promises of future rewards for years of loyalty now. But it has not actually brought us into greater awareness of what is going on around us. We are not approaching some Zen state of an infinite moment, completely at one with our surroundings, connected to others, and aware of ourselves on any fundamental level.

Rather, we tend to exist in a distracted present, where forces on the periphery are magnified and those immediately before us are ignored. Our ability to create a plan—much less follow through on it—is undermined by our need to be able to improvise our way through any number of external impacts that stand to derail us at any moment. Instead of finding a stable foothold in the here and now, we end up reacting to the ever-present assault of simultaneous impulses and commands.

In some senses, this was the goal of those who developed the computers and networks on which we depend today. Mid-twentieth-century computing visionaries Vannevar Bush and J. C. R. Licklider dreamed of developing machines that could do our remembering for us. Computers would free us from the tyranny of the past—as well as the horrors of World War II—allowing us to forget everything and devote our minds to solving the problems of today. The information would still be there; it would simply be stored out of body, in a machine.

It's a tribute to both their designs on the future and their devotion to the past that they succeeded in their quest to free up the present of the burden of memory. We have, in a sense, been allowed to dedicate much more of our cognitive resources to active RAM than to maintaining our cerebral-storage hard drives. But we are also in danger of squandering this cognitive surplus on the trivial pursuit of the immediately relevant over any continuance of the innovation that got us to this point.

Behavioral economists exploit the growing disparity between our understanding of the present and that of the future, helping us see future debts as less relevant than current costs and leading us to make financial decisions against our own better interests. As these ways of understanding debt and lending trickle up to those making decisions about banking and macrofinance—such as the Federal Reserve or the European Central Bank—our greater economies end up suffering from the same sorts of logical traps as those of individual mortgage holders and credit card users.

Neuroscientists, mostly at the service of corporations looking to develop more compliant employees and consumers, are homing in on the way people make choices. But no matter how many subjects they put in their MRI machines, the focus of this research is decision making in the moment, the impulsive choices made in the blink of an eye, rather than those made by the lobes responsible for rational thought or consideration. By implementing their wares solely on the impulsive—while diminishing or altogether disregarding the considered—they push us toward acting in what is thought of as an instinctual, reptilian fashion.

And this mode of behavior is then justified as somehow more

connected to the organic, emotional, and immediately relevant moment in which human beings actually live. Of course, this depiction of consciousness may help sell the services of neurotechnicians to advertisers, but it does not accurately represent how the human brain relates to the moment in which the organism exists.

No matter how invasive the technologies at their disposal, marketers and pollsters never come to terms with the living process through which people choose products or candidates; they are looking at what people just bought or thought, and making calculations based on that after-the-fact data. The "now" they seek to understand tells them nothing about desire, reasons, or context. It is simply an effort to key off what we have just done in order to manipulate our decisions in the future. Their campaigns encourage the kinds of impulsive behavior that fool us into thinking we are living in the now while actually just making us better targets for their techniques.

That is because there is no *now*—not the one they're talking about, anyway. It is necessarily and essentially trivial. The minute the "now" is apprehended, it has already passed. Like they used to say about getting one's picture on a *Time* magazine cover: the moment something is realized, it is over. And like the diminishing beauty returns for a facially paralyzed Botox addict, the more forcefully we attempt to stop the passage of time, the less available we are to the very moment we seek to preserve.

As a result, our culture becomes an entropic, static hum of everybody trying to capture the slipping moment. Narrativity and goals are surrendered to a skewed notion of the real and the immediate; the Tweet; the status update. What we are doing at any given moment becomes all-important—which is behavioristically doomed. For this desperate approach to time is at once flawed and narcissistic. Which "now" is important: the now I just lived or the now I'm in right now?

In the following chapters, we will explore present shock as it manifests in a variety of ways, on a myriad of levels. We will look at how it changes the way we make and experience culture, run our businesses, invest our money, conduct our politics, understand science, and make sense of our world. In doing so, we will consider panic reactions to present shock right alongside more successful approaches to living outside what we have always thought of as time.

The book is divided into five sections, corresponding to the five main ways that present shock manifests for us. We begin with the collapse of narrative. How do we tell stories and convey values without the time required to tell a linear story? How does pop culture continue to function without traditional storylines, and how does politics communicate without grand narratives? We move on to "Digiphrenia"—the way our media and technologies encourage us to be in more than one place at the same time. We'll see that our relationship to time has always been defined by the technologies we use to measure it, and that digital time presents particular challenges we haven't had to contend with before. In "Overwinding," we look at the effort to squish really big timescales into much smaller ones. It's the effort to make the passing moment responsible for the sorts of effects that actually take real time to occur. In particular, what does this do to business and finance, which are relying on increasingly derivative forms of investment? Next we look at what happens when we try to make sense of our world entirely in the present tense. Without a timeline through which to parse causes and effects, we instead attempt to draw connections from one thing to another in the frozen moment, even when such connections are forced or imaginary. It's a desperate grasp for real-time pattern recognition I'll call "Fractalnoia." Finally, we face "Apocalypto"—the way a seemingly infinite present makes us long for endings, by almost any means necessary.

We will encounter drone pilots contending with the stress of dropping bombs on a distant war zone by remote control before driving home to the suburbs for supper an hour later. We will see the way the physical real estate of Manhattan is being optimized for the functioning of the ultrafast trading algorithms now running the stock market—as well as what this means for the human traders left in the wake. We will encounter doomsday "preppers" who stock up on silver coins and ready-to-eat meals while dismissing climate change as a conspiracy theory hatched by Al Gore and since exposed in an email scandal.[3] We will consider the "singularity"—as well as our scientific community's response to present shock—especially for the ways it mirrors the religious extremism accompanying other great social shifts throughout history.

Most important, we will consider what we human beings can do to pace ourselves and our expectations when there's no temporal backdrop against which to measure our progress, no narrative through which to make sense of our actions, no future toward which we may strive, and seemingly no time to figure any of this out.

I suggest we intervene on our own behalf—and that we do it right now, in the present moment. When things begin accelerating wildly out of control, sometimes patience is the only answer. Press pause.

We have time for this.

CHAPTER 1

NARRATIVE
COLLAPSE

I had been looking forward to the twenty-first century.

That's what most of us were doing in the 1990s: looking forward. Everything seemed to be accelerating, from the speed of technology to the growth of markets. PowerPoint presentations everywhere used the same steep upward curve to describe the way business revenues, computer use, carbon dioxide emissions, and growth of every kind were accelerating exponentially.

Moore's Law, a rule of thumb for technological progress coined in 1965 by Intel cofounder Gordon E. Moore, told us that our computer-processing speeds would double about every two years. Along with that, however, everything else seemed to be doubling as well—our stock indexes, medical bills, Internet speeds, cable-TV stations, and social networks. We were no longer adjusting to individual changes, we were told, but to the accelerating rate of change itself. We were in what futurist Alvin Toffler called "future shock."

As a result, everything and everyone was leaning toward the future. We weren't looking forward to anything in particular so much as we were simply looking forward. Trend casters and "cool hunters" became the highest-paid consultants around, promising exclusive peeks at what lie ahead. Optimistic books with titles like "The Future of This" or "The Future of That" filled the store shelves, eventually superseded by pessimistic ones titled "The End of This" or "The End of That." The subjects themselves mattered less than the fact that they all either had a future or—almost more reassuringly— did not.

We were all futurists, energized by new technologies, new theories, new business models, and new approaches that promised not just more of the same, but something different: a shift of an uncertain nature, but certainly of unprecedented magnitude. With each passing year, we seemed to be closer to some sort of chaos attractor that was beckoning us toward itself. And the closer we got, the more time itself seemed to be speeding up. Remember, these were the last years of the last decade of the last century of the millennium. The roaring, net-amplified, long boom of the 1990s seemed defined by this leaning forward, this ache toward conclusion, this push toward 2000 and the ultimate calendar flip into the next millennium.

Though technically still in the twentieth century, the year 2000 was a good enough marker to stand in for millennial transformation. So we anticipated the change like messianic cultists preparing for the second coming. For most of us, it took the less religious form of anticipating a Y2K computer bug where systems that had always registered years with just two digits would prove incapable of rolling over to 00. Elevators would stop, planes would fall out of the sky, nuclear plants would cease to cool their reactor cores, and the world as we know it would end.

Of course, if the changeover didn't get us, the terrorists would. The events of 9/11 hadn't even happened yet, but on the evening of

December 31, 1999, Americans were already on alert for a violent disruption of the Times Square New Year's Eve festivities. Seattle had canceled its celebration altogether, in anticipation of an attack. CNN's coverage circled the globe from one time zone to another as each hit midnight and compared the fireworks spectacle over the Eiffel Tower to the one at the Statue of Liberty. But the more truly spectacular news reported at each stop along the way that night was that nothing spectacular happened at all. Not in Auckland, Hong Kong, Cairo, Vatican City, London, Buenos Aires, or Los Angeles. The planes stayed in the sky (all but three of KLM's 125-plane fleet had been grounded just in case), and not a single terror incident was reported. It was the anticlimax of the millennium.

But something did shift that night as we went from years with 19's to those with 20's. All the looking forward slowed down. The leaning into the future became more of standing up into the present. People stopped thinking about where things were going and started to consider where things were.

In the financial world, for example, an investment's future value began to matter less than its current value. Just ten weeks into the millennium, the major exchanges were peaking with the tech-heavy and future-focused NASDAQ reaching its all-time high, over 5,100 points. Then the markets started down—and have never quite recovered. Although this was blamed on the dot.com bubble, the market's softening had nothing to do with digital technologies actually working (or not) and everything to do with a larger societal shift away from future expectations and instead toward current value. When people stop looking to the future, they start looking at the present. Investments begin to matter less for what they might someday be worth, because people are no longer thinking so much about "someday" as they are about today. A stock's "story"—the rationale for why it is going to go up—begins to matter less than its actual value in *real* time. What are my stocks worth as of this moment? What do I really own? What is the value of my portfolio right *now*?

The stock market's infinite expansion was just one of many stories dependent on our being such a future-focused culture. All the great "isms" of the twentieth century—from capitalism to communism to Protestantism to republicanism to utopianism to messianism—depended on big stories to keep them going. None of them were supposed to be so effective in the short term or the present. They all promised something better in the future for having suffered through something not so great today. (Or at least they offered something better today than whatever pain and suffering supposedly went on back in the day.) The ends justified the means. Today's war was tomorrow's liberation. Today's suffering was tomorrow's salvation. Today's work was tomorrow's reward.

These stories functioned for quite a while. In the United States, in particular, optimism and a focus on the future seemed to define our national character. Immigrants committed to a better tomorrow risked their lives to sail the ocean to settle a wilderness. The New World called for a new story to be written, and that story provided us with the forward momentum required to live for the future. The Protestant work ethic of striving now for a better tomorrow took hold in America more powerfully than elsewhere, in part because of the continent's ample untapped resources and sense of boundless horizon. While Europe maintained the museums and cultures of the past, America thought of itself as forging the new frontier.

By the end of World War II, this became quite true. Only, America's frontier was less about finding new territory to exploit than it was about inventing new technologies, new businesses, and new ideas to keep the economy expanding and the story unfolding. Just as Mormonism continued the ancient story of the Bible into the American present, technologies, from rocket ships to computer chips, would carry the story of America's manifest destiny into the future. The American Dream, varied though it may have been, was almost universally depending on the same greater shape, the same kind of story to carry us along. We were sustained economically, politically, and even spiritually, by stories.

Together these stories helped us construct a narrative experience of our lives, our nation, our culture, and our faith. We adopted an entirely storylike way of experiencing and talking about the world. Through the lens of narrative, America isn't just a place where we live but is a journey of a people through time. Apple isn't a smart phone manufacturer, but two guys in a garage who had a dream about how creative people may someday gain command over technology. Democracy is not a methodology for governing, but the force that will liberate humanity. Pollution is not an ongoing responsibility of industry, but the impending catastrophic climax of human civilization.

Storytelling became an acknowledged cultural value in itself. In front of millions of rapt television viewers, mythologist Joseph Campbell taught PBS's Bill Moyers how stories provide the fundamental architecture for human civilization. These broadcasts on *The Power of Myth* inspired filmmakers, admen, and management theorists alike to incorporate the tenets of good storytelling into their most basic frameworks. Even brain scientists came to agree that narrativity amounted to an essential component of cognitive organization. As Case Western Reserve University researcher Mark Turner concluded: "Narrative imagining—story—is the fundamental instrument of thought. Rational capacities depend upon it. It is our chief means of looking into the future, of predicting, of planning, and of explaining."[1] Or as science fiction writer Ursula K. Le Guin observed, "The story—from Rapunzel to *War and Peace*—is one of the basic tools invented by the human mind, for the purpose of gaining understanding. There have been great societies that did not use the wheel, but there have been no societies that did not tell stories."[2]

Experiencing the world as a series of stories helps create a sense of context. It is comforting and orienting. It helps smooth out obstacles and impediments by recasting them as bumps along the way to some better place—or at least an end to the journey. As long as there's enough momentum, enough forward pull, and enough dramatic tension, we can suspend our disbelief enough to stay in the story.

The end of the twentieth century certainly gave us enough momentum, pull, and tension. Maybe too much. Back in the quaint midcentury year of 1965, *Mary Poppins* was awarded five Oscars, the Grateful Dead played their first concert, and *I Dream of Jeannie* premiered on NBC. But it was also the year of the first spacewalk, the invention of hypertext, and the first successful use of the human respirator. These events and inventions, and others, were promising so much change, so fast, that Alvin Toffler was motivated to write his seminal essay "The Future as a Way of Life," in which he coined the term "future shock":

> We can anticipate volcanic dislocations, twists and reversals, not merely in our social structure, but also in our hierarchy of values and in the way individuals perceive and conceive reality. Such massive changes, coming with increasing velocity, will disorient, bewilder, and crush many people. . . . Even the most educated people today operate on the assumption that society is relatively static. At best they attempt to plan by making simple straight-line projects of present-day trends. The result is unreadiness to meet the future when it arrives. In short, future shock.[3]

Toffler believed things were changing so fast that we would soon lose the ability to adapt. New drugs would make us live longer; new medical techniques would allow us to alter our bodies and genetic makeup; new technologies could make work obsolete and communication instantaneous. Like immigrants to a new country experiencing culture shock, we would soon be in a state of future shock, waking up in a world changing so rapidly as to be unrecognizable. Our disorientation would have less to do with any particular change than the rate of change itself.

So Toffler recommended we all become futurists. He wanted kids to be taught more science fiction in school, as well as for them to take special courses in "how to predict." The lack of basic predictive skills would for Toffler amount to "a form of functional illiteracy in the contemporary world."[4]

To a great extent this is what happened. We didn't get futurism classes in elementary school, but we did get an abject lesson in futurism from our popular and business cultures. We all became futurists in one way or another, peering around the corner for the next big thing, and the next one after that. But then we actually got there. Here. Now. We arrived in the future. That's when the story really fell apart, and we began experiencing our first true symptoms of present shock.

NARRATIVE COLLAPSE

Toffler understood how our knowledge of history helps us put the present in perspective. We understand where we are, in part, because we have a story that explains how we got here. We do not have great skill in projecting that narrative ability into the future. As change accelerated, this inability would become a greater liability. The new inventions and phenomena that were popping up all around us just didn't fit into the stories we were using to understand our circumstances. How does the current story of career and retirement adjust to life spans increasing from the sixties to the one hundreds? How do fertility drugs change the timeline of motherhood, how does email change our conception of the workweek, and how do robots change the story of the relationship of labor to management? Or, in our current frame of reference, how does social networking change the goals of a revolution?

If we could only get better at imagining scenarios, modeling future realities, and anticipating new trends, thought Toffler, we may be less traumatized by all the change. We would be equipped to imagine new narrative pathways that accommodated all the disruptions.

Still, while *Star Trek* may have correctly predicted the advent of cell phones and iPads, there are problems inherent to using science fiction stories to imagine the future. First, sometimes reality moves even faster and less predictably than fiction. While stories must follow certain plot conventions in order to make sense to their audiences, reality is under no such obligation. Stuff just happens, and rarely on schedule. Second, and more significant, stories are usually much less about predicting the future than influencing it. As a medium, stories have proven themselves great as a way of storing information and values, and then passing them on to future generations. Our children demand we tell them stories before they go to bed, so we lace those narratives with the values we want them to take with them into their dreams and their adult lives. Likewise, the stories and myths of our religions and national histories preserve and promote certain values over time. That's one reason civilizations and their values can persist over centuries.

The craft of futurism—however well intentioned—almost always comes with an agenda. For those who were already familiar with the Internet, the first issues of *Wired* magazine seemed glaringly obvious in their underlying purpose to marry the values of the net with those of the free market. The many futurists who emerged in the late 1990s simply couldn't help but predict futures in which the most important specialists to have around would be—you guessed it—futurists. The stories they came up with were tailor-made for corporations looking for visions of tomorrow that included the perpetuation of corporate power. Futurism became less about predicting the future than pandering to those who sought to maintain an expired past.

Meanwhile, all this focus on the future did not do much for our ability to contend with the present. As we obsessed over the future of this and the future of that, we ended up robbing the present of its ability to contribute value and meaning. Companies spent more money and energy on scenario planning than on basic competency. They hired consultants (sometimes media theorists, like me) to give them "mile-high views" on their industries. The higher up they could go, they imagined, the farther ahead they could see. One technology company I spoke with was using research and speculation on currency futures to decide where to locate offshore factories. The CFO of another was busy hedging supply costs by betting on commodities futures—with little regard to emerging technologies in his own company that would render the need for such commodities obsolete. Some companies lost millions, or even went out of business, making bets of this sort on the future while their core competencies and innovative capabilities withered.

As people, businesses, institutions, and nations, we could maintain our story of the future only by wearing increasingly restrictive blinders to block out the present. Business became strategy, career became a route to retirement, and global collaboration became brinksmanship. This all worked as long as we could focus on those charts where everything pointed up. But then the millennium actually came. And then the stock market crashed. And then down came the World Trade Towers, and the story really and truly broke.

The discontinuity generated by the 9/11 attacks should not be underestimated. While I was writing this very chapter, I met with a recent college graduate who was developing a nonprofit company and website to help create relationships between "millennials" of her generation and more aged mentors of my own. She explained that her generation was idealistic enough to want to help fix the world, but that they had been "traumatized by 9/11 and now we're incapable of accessing the greater human projects." Somehow, she felt, the tragedy had disconnected her generation from a sense of history and purpose, and that they "needed to connect with people from before that break in the story in order to get back on track."

This was also the generation who used their first access to the polls to vote for Obama. She and her friends had supported his campaign and responded to his explicitly postnarrative refrain, borrowed from Alice Walker's book title: "We are the ones we have been waiting for. We are the change we seek." What a call to presentism this was! Young people took Obama at his word, rising to the challenge to become change rather than wait for it. Of course, it turned out to be more of a campaign slogan than an invitation to civic participation—just more rhetoric for a quite-storybook, ends-justify-the-means push to power. It would be left to the Occupy movement to attempt a genuinely presentist approach to social and political change. But Obama's speechwriters had at least identified the shift under way, the failure of stories to create a greater sense of continuity, and the growing sense that something much more immediate and relevant needed to take their place.

BIG STORIES

Traditional stories, with traditional, linear arcs, have been around for a long time because they work. They seem to imitate the shape of real life, from birth to death. Like a breath or lovemaking, these sorts of stories have a rise and a satisfying fall; a beginning, a middle, and an end. While it seems quite natural to us today, this familiar shape didn't become the default structure of stories until pretty late in human history, after the invention of text and scrolls, in literate cultures such as ancient Greece.

The Bible's stories—at least the Old Testament's—don't work quite the same way. They were based more in the oral tradition, where the main object of the storyteller was simply to keep people involved in the moment. Information and morals were conveyed, but usually by contrasting two characters or nations with one another— one blessed, the other damned. Epic poems and, later, theater, followed the more linear progression we might better associate with a scroll or bound book. There's a beginning and there's an end. Wherever we are in the story, we are aware that there are pages preceding and pages to come. Our place in the scroll or book indicates how close we are to finishing, and our emotional experience is entirely bound up in time.

Aristotle was the first, but certainly not the last, to identify the main parts of this kind of story, and he analyzed them as if he were a hacker reverse-engineering the function of a computer program. The story mechanics he discovered are very important for us to understand, as they are still in use by governments, corporations, religions, and educators today as they attempt to teach us and influence our behaviors. They are all the more important for the way they have ceased to work on members of a society who have gained the ability to resist their spell. This has put the storytellers into present shock.

The traditional linear story works by creating a character we can identify with, putting that character in danger, and then allowing him or her to discover a way out. We meet Oedipus, Luke Skywalker, or Dora the Explorer. Something happens—an initiating event—that sends the character on a quest. Oedipus wants to find the truth of his origins; Luke wants to rescue Princess Leia; Dora wants to get the baby frog back into its tree. So then the character makes a series of choices that propel him or her into increasingly dangerous situations. Oedipus decides to find and kill the murderer of King Laius; Luke becomes a Jedi to fight the Empire; Dora enlists her monkey pal, Boots, to help her bring the baby frog through

the scary forest to its home. At each step along the way, the character proceeds further into peril and takes the audience further up the
path into tension and suspense.

Just when the audience has reached its peak of anxiety—the
place where we can't take any more without running out of the theater or throwing the book on the floor—we get our reversal. Oedipus learns that the murderer he seeks is himself; Luke learns that
Darth Vader is his father; Dora learns she herself holds the answer
to the ugly old troll's riddle. And with that, finally, comes full recognition and release of tension. Oedipus blinds himself, Luke brings
his dying father back to the light side of the force, and Dora gets the
baby frog to its family's tree. Most important, the audience gets catharsis and relief. The ride is over. The greater the tension we were
made to tolerate, the higher up the slope we get, and the more we
can enjoy the way down.

This way of organizing stories—Joseph Campbell's "heroic journey"[5]—is now our way of understanding the world. This may have
happened because the linear structure is essentially true to life, or we
may simply have gotten so accustomed to it that it now informs the
way we look at events and problems that emerge. Whatever the case,
this structure also worked perfectly for conveying values of almost
any kind to the captivated audience. For if we have followed the protagonist into danger, followed him up the incline plane of tension
into a state of great suspense and anxiety, we will be willing to accept whatever solution he is offered to get out. Arnold Schwarzenegger
finds a new weapon capable of killing the bad aliens, the interroga-
tor on *Law & Order* uses psychology to leverage the serial killer's ego
against himself, or the kids on *Glee* learn that their friendships matter more than winning a singing contest. The higher into tension
we have gone, the more dependent we are on the storyteller for a
way out. That's why he can plug in whatever value, idea, or moral
he chooses.

Or product. The technique reaches its height, of course, in any typical television commercial. In just thirty seconds (or twenty-eight seconds, when you account for the fades to and from video blackness), a character finds himself in a situation, makes choices that put him in danger, and then finds a solution in the form of a purchase. For just one actual example: A girl is anticipating her high school prom when she notices a pimple on her cheek (initiating event). She tries hot compresses, popping it, and home remedies, which only make it worse (rising tension). Just when it looks as though there's no way to avoid being terribly embarrassed and humiliated at her prom, a friend sees the pimple and, instead of teasing her, tells her about the new fast-acting pimple cream (reversal). She puts on the cream (recognition) and goes to the prom, pimple free (catharsis).

If we have followed the character up the ramp of tension into danger, then we must swallow the pill, cream, gun, or moral the storyteller uses to solve the problem. For all this to work, however, the storyteller is depending on a captive audience. The word "entertainment" literally means "to hold within," or to keep someone in a certain frame of mind. And at least until recently, entertainment did just this, and traditional media viewers could be depended on to sit through their programming and then accept their acne cream.

Even if television viewers sensed they were being drawn into an anxious state by a storytelling advertiser who simply wanted to push a product, what were the alternatives? Before the advent of interactive devices like the remote control, the television viewer would have had to get up off the couch, walk over to the television set, turn the dial, tune in the new station, and then adjust the rabbit ears. Or simply walk out of the room and possibly miss the first moments of the show when the commercial ended. Although television viewers weren't as coerced into submission as a churchgoer forced to stay in the pew and listen to the story as the minister related it, they were still pretty much stuck swallowing whatever pill the programmer inserted into the turning point of the narrative.

Then came interactivity. Perhaps more than any postmodern

idea or media educator, the remote control changed the way we related to television, its commercials, and the story structure on which both depended. Previously, leaving the couch and walking up to the television to change the channel might cost more effort than merely enduring the awful advertisement and associated anxiety. But with a remote in hand, the viewer can click a button and move away effortlessly. Add cable television and the ability to change channels without retuning the set (not to mention hundreds of channels to watch instead of just three), and the audience's orientation to the program has utterly changed. The child armed with the remote control is no longer watching a television program, but watching television—moving away from anxiety states and into more pleasurable ones.

Take note of yourself as you operate a remote control. You don't click the channel button because you are bored, but because you are mad: Someone you don't trust is attempting to make you anxious. You understand that it is an advertiser trying to make you feel bad about your hair (or lack of it), your relationship, or your current SSRI medication, and you click away in anger. Or you simply refuse to be dragged still further into a comedy or drama when the protagonist makes just too many poor decisions. Your tolerance for his complications goes down as your ability to escape becomes increasingly easy. And so today's television viewer moves from show to show, capturing important moments on the fly. Surf away from the science fiction show's long commercial break to catch the end of the basketball game's second quarter, make it over to the first important murder on the cop show, and then back to the science fiction show before the aliens show up.

Deconstructed in this fashion, television loses its ability to tell stories over time. It's as if the linear narrative structure had been so misused and abused by television's incompetent or manipulative storytellers that it simply stopped working, particularly on younger people who were raised in the more interactive media environment and equipped with defensive technologies. And so the content of television, and the greater popular culture it leads, adapted to the new situation.

NOW-IST POP CULTURE IS BORN

Without the time or permission to tell a linear story with a beginning, a middle, and an end, television programmers had to work with what they had—the moment. To parents, educators, and concerned experts, the media that came out of this effort has looked like anything but progress. As Aristotle explained, "When the storytelling in a culture goes bad the result is decadence."[6] At least on a surface level, the new storyless TV shows appeared to support Aristotle's maxim.

Animated shows masquerading as kids' programming, such as *Beavis and Butt-head* (1993) and *The Simpsons* (1989), were some of the first to speak directly to the channel surfer.[7] MTV's animated hit *Beavis and Butt-head* consists of little more than two young teenagers sitting on a couch watching MTV rock videos. Though dangerously mindless in the view of most parents, the show artfully recapitulates the experience of kids watching MTV. As the two knuckleheads comment on the music videos, they keep audience members aware of their own relationship to MTV imagery. The show takes the form of a screen-within-a-screen, within which typical MTV videos play. But where a rock video may normally entice the viewer with provocative or sexual imagery, now the viewer is denied or even punished for being drawn in. When the sexy girl comes into frame, Butt-head blurts out, "Nice set"; Beavis giggles along; and the viewer is alienated from the imagery. The two animated kids are delivering a simple object lesson in media manipulation. When they don't like something, one says, "This sucks, change it," and the other hits the remote. Beavis and Butt-head might not have singlehandedly rendered the rock video obsolete, but their satire provided a layer of distance and safety between viewers and the programming they no longer trusted.

The cult hit *Mystery Science Theater 3000* (first aired in 1988) turned this genre into something close to an art form. Set in the future, the show allows the audience to watch along as the sole captive inhabitant of a space station and his two robot companions are forced to view bad B movies and low-budget science fiction sagas. Our television screen shows us the movie, with the heads of the three audience members in silhouette, Looney Tunes–style, several rows ahead of us. The trio makes comments and wisecracks about the movie, much as we may do if we were watching with our own friends.

But we're not. For the most part, viewers of this late-night show are isolated in their apartments, using the images on their screens as surrogate companions. In a self-similar fashion, the character trapped in the futuristic space station has fashioned his own robot friends out of spare projection parts—the ones that could have given him some control over when the movies are shown. He uses the technology at his disposal to provide himself with simulated human interaction but has given up a certain amount of freedom to do so. So, too, do the young viewers of the show simulate a social setting with their television sets, suffering through the long, awful sci-fi movies delivered on the network's schedule for the joy of simulated companionship. *MST3K*, as its fans call it, is both entertainment and mirror. If we can no longer follow a character through his story over time, we can instead *be* that character in the moment. Most of the film's dialogue is drowned out by the antics in the audience, and the plot is lost to the endless succession of jokes and mimicry. The linear progression of the film's story is sacrificed to the more pressing need for a framework that mirrors the viewing experience.

The individual jokes and asides of the characters also make up a new media education for the show's audience. Almost all of the humor is derived from references to other media. The robots make an Andrew Lloyd Webber grill to burn the composer's self-derivative scores and argue about the relative merits of the Windows and Macintosh operating systems. When they observe Bela Lugosi taking off his lab coat in a campy old sci-fi feature, the robots sing, "It's a beautiful day in the laboratory" to the tune of the *Mister Rogers' Neighborhood* theme. The robots make sure to call attention to every cheesy special effect and structural flaw. As the noise of guns and guard dogs pursue escaping convicts, a robot shouts, "Sounds like the foley artists are chasing us. Move it!" Toward the end of another film, one robot comments, "Isn't this a little late in act three for a plot twist?"

To appreciate the humor of the show, viewers need to understand the media as a self-reflexive universe of references, any of which can be used to elucidate any other. Each joke is a demonstration of the media's self-similarity. This is not a humor of random association but a comedy of connectivity where images and ideas from very disparate sources are revealed as somehow relevant to one another. To belong to the *MST3K* culture is to understand at least some of the literally hundreds of references per show and, more important, how they relate to one another. When this is not the object of the game, the characters instead keep their audience aware of their moment-to-moment relationship to the media, either by commenting on the technical quality of the film or by calling attention to themselves as recapitulated bracketing devices.

The Simpsons, now in its twenty-fourth season of self-referential antics, brings the same TV-within-a-TV sensibility to an even wider, mainstream audience. The opening theme still plays over animation of the entire family rushing home to the living room couch in time for their favorite show. Mirroring our increasingly ironic sensibility, the program's child protagonist, Bart Simpson, seems aware of his own role within the show and often comments on what his family must look like to the audience watching along.

Although *The Simpsons* episodes have stories, these never seem to be the point. There are no stakes: characters die, or do things that would kill them, yet reappear in later episodes. The fact that Homer (after the Greek poet) Simpson might have caused a nuclear spill does not create tension in the typical sense, and nobody watching particularly cares whether the town of Springfield is spared the resulting devastation. We are not in a state of suspense. Instead, the equivalents of recognition or reversal come from recognizing what other forms of media are being satirized in any given moment. When Homer picks up his daughter from child care, she is perched on a wall next to hundreds of other pacifier-sucking babies. The "a-ha" moment comes from recognizing it is a spoof of Hitchcock's *The Birds*—and that institutional child care has taken on the quality of a horror movie. Unlike the heroes depicted by his namesake, Homer has no epic journey. He remains in a suspended, infinite present, while his audience has all the recognitions.

Still on the air after all these years, *The Simpsons*, along with the many satirical, self-referential shows that followed its path (the creators of *Family Guy*, *South Park*, and even *The Office* all credit *The Simpsons* as a seminal influence), offers the narrative-wary viewer some of the satisfaction that traditional stories used to provide—but through nonnarrative means. *Family Guy* (1999), canceled by FOX in 2002 but revived in 2005 when its popularity online kept growing, seems tailor-made for the YouTube audience. The show's gags don't even relate to the story or throughline (such as they are), but serve as detours that thwart or halt forward motion altogether. Rather than simply scripting pop culture references into the scenes, *Family Guy* uses these references more as wormholes through which to escape from the temporal reality of the show altogether—often for minutes at a time, which is an eternity on prime-time television. In one episode the mom asks her son to grab a carton of milk, "and be sure to take it from the back." Apropos of nothing, a black-and-white sketch of a man's hand pulls the child into an alternate universe of a-ha's iconic 1984 "Take On Me" music video. The child runs through a paper labyrinth with the band's front man for the better part of a minute before suddenly breaking through a wall and back into the *Family Guy* universe.

This reliance on what the show's YouTube fans call "cutscenes" turns what would have been a cartoon sitcom into a sequence of infinite loops, each one just as at home on the decontextualized Internet as they are strung together into a half hour of TV. The only real advantage to watching them in their original form within the program is the opportunity to delight in the writers' audacious disregard for narrative continuity (and for pop culture as a whole).

Finally, going so far out on this postnarrative journey that it comes full circle, NBC's unlikely hit *Community* (2009) is ostensibly a plotted sitcom about a group of misfits at Greendale Community College—except for the fact that the characters continually refer to the fact that they are on a television sitcom. For example, as Greendale's principal completes his standard PA announcements at the opening of one scene, the character Abed—a pop-culture-obsessed voyeur with Asperger's syndrome who is often a proxy for the audience—remarks that the announcement "makes every ten minutes feel like the beginning of a new scene of a TV show." He continues, "Of course, the illusion only lasts until someone says something they'd never say on TV, like how much their life is like TV. There, it's gone."

Community assumes such extensive pop cultural literacy that even its narrative tropes—odd couple turned best friends; triumph of the underdog; will they or won't they *do* it?—are executed with

dripping irony. These are overwrenched plots, recognized as parody
by an audience well versed in television's all too familiar narrative
arcs. They even do one of those highlights episodes stringing together scenes from previous episodes (the kind that normal sitcoms
do to fill up an episode with old free footage), except none of the
scenes are actually from previous episodes. It's a series of fake flashbacks to scenes that never appeared in those episodes—a satire of the
clip-show trope. "*Community*," writes Hampton Stevens in the *Atlantic*, "isn't actually a sitcom—any more than *The Onion* is an actual
news-gathering organization. *Community*, instead, is a weekly satire of
the sitcom genre, a spoof of pop culture in general."[8] While *The Simpsons* and *Family Guy* disrupt narrative in order to make pop culture
references, *Community*'s stories are themselves pop culture references.
Narrative becomes a self-conscious wink.

Through whichever form of postmodern pyrotechnics they practice, these programs attack the very institutions that have abused
narrative to this point: advertisers, government, religions, pop culture sellouts, politicians, and even TV shows themselves. They don't
work their magic through a linear plot, but instead create contrasts
through association, by nesting screens within screens, and by giving viewers the tools to make connections between various forms of
media. It's less like being walked along a pathway than it is like
being taken up high and shown a map. The beginning, the middle,
and the end have almost no meaning. The gist is experienced in
each moment as new connections are made and false stories are exposed or reframed. In short, these sorts of shows teach pattern recognition, and they do it in real time.

Of course, this self-conscious parody was just one of many responses to a deconstructing mediascape. TV and movies, low culture and high culture, have all been contending with the collapse of
narrative. Some resist and some actively contribute; some complain
while others celebrate. We are just now finding a new equilibrium

in a transition that has taken over twenty years—most visibly in the cinema. As if responding to the disruption of the remote control and other deconstructive tools and attitudes, many American films of the late 1990s seemed to be searching for ways to preserve the narrative structure on which their messages and box office receipts were depending.

Movies dedicated to preserving the stories we use to understand ourselves turned the cut-and-paste technologies against the digital era from which they emerged, as if to restore the seamless reality of yesterday. The mid-1990s blockbuster *Forrest Gump*, for just one example, attempted to counteract the emerging discontinuity of the Internet age by retelling the story of the twentieth century from the perspective of a simpleton. Filmmaker Robert Zemeckis was already most famous for the *Back to the Future* series in which his characters went back in time to rewrite history. *Forrest Gump* attempts this same revisionist magic through a series of flashbacks, in which the audience relives disjointed moments of the past century of televised history, all with Gump magically pasted into the frame. We see Gump protesting the Vietnam War, Gump with John Lennon, and even Gump meeting JFK and saying he needs to pee.

Gump's lack of awareness allows him to fall, by sheer luck, into good fortune at every turn. He becomes a war hero and multimillionaire by blindly stumbling through life with nothing more than the good morals his mom taught him, while the people around him who seem more aware of their circumstances drop like flies from war wounds, AIDS, and other disasters. In this story's traditionally narrative schema, Gump is saved and most everyone else is damned. The impending unpredictability of life beyond narrative is reinterpreted as a box of chocolates—"You never know what you're gonna get." But it's a box of chocolates! You can pretty well count on getting a chocolate as long as you don't reach outside of the box into the real world of sharp rocks and biting bugs. The opening sequence of the movie tells it all: in one continuous shot a feather floats on the wind, effortlessly wandering over the rooftops of a small, perfect town, and lands at Gump's feet, either coincidentally or by divine will. Of course, it was neither luck nor God's guiding the feather's path, but the will of the movie's director, who used cinematic trickery to create the continuous sequence. Just like Gump, we, the audience, are kept ignorant of the special effects, edits, and superimpositions, as technology is exploited to make the facade look seamless and real. And what does Gump do with the feather? He puts it in an old box with his other collected trinkets—contained, like everything else, within his oversimplified narrative.

If *Forrest Gump* could be considered a defender of the narrative worldview, its mid-1990s contemporary, Quentin Tarantino's *Pulp Fiction*, may be thought of as its opposite. Where Gump offers us a linear, if rewritten, historical journey through the decades since World War II, *Pulp Fiction* compresses imagery from those same years into a stylistic pastiche. Every scene has elements from almost every decade—a 1940s-style suit, a 1950s car, a 1970s telephone, a 1990s retro nightclub—forcing the audience to give up its attachment to linear history and accept instead a vision of American culture as a compression of a multitude of eras, and those eras themselves being reducible to iconography as simple as a leather jacket or dance step. The narrative technique of the film also demands that its audience abandon the easy plot tracking offered by sequential storytelling. Scenes occur out of order and dead characters reappear. On one level we are confused; on another, we are made privy to new kinds of information and meaning. The reordering of sequential events allows us to relate formerly nonadjacent moments of the story to one another in ways we couldn't if they had been ordered in linear fashion. If we watch someone commit a murder in one scene, our confusion about his motivations can be answered by going backward in time in the very next scene. The movie's final protagonist, Bruce Willis, comically risks his life to retrieve the single heirloom left to him by his father: his watch. *Pulp Fiction* delights in its ability to play with time, and in doing so shows us the benefits of succumbing to the chaos of a postnarrative world. The object of the game is to avoid getting freaked out by the resulting gaps, juxtapositions, and discontinuity.

Slowly but surely, dramatic television and cinema seemed to give up the fight, and instead embrace the timelessness, even the purposelessness, of living in the present. The classic situation comedy had been narrative in its construction. The "situation" usually consisted of a history so important to the show that it was retold during the opening theme song. A poor mountaineer was shooting at some food, accidentally struck oil, got rich, and brought his whole hillbilly family to Beverly Hills. A three-hour boat tour meets with a storm at sea, shipwrecking a group of unlikely castaways. Compared to these setups, modern sitcoms appear as timelessly ahistorical as *Waiting for Godot*. *Friends* chronicles the exploits of some people who happen to frequent the same coffee bar. *Seinfeld* is a show about nothing. The backstory of *Two and a Half Men* has more to do with Charlie Sheen's dismissal and Twitter exploits than the divorces of the show's in-world characters. These shows are characterized by their frozenness in time, as well as by the utter lack of traditional narrative goals.

The new challenge for writers is to generate the sense of captivity, as well as the sensations and insights, of traditional narrative—but to do so without the luxury of a traditional storyline. So they come up with characters who simply wake up in a situation and have to figure out who they are or what the heck is going on around them. The characters are contending with the same present shock as their creators.

The movie *Memento* follows a man who loses his memory every five minutes or so and must repiece together his existence (and a murder mystery) in essentially no time at all. He tattoos clues and insights to his body, turning himself into a mosaic of hints. If he is able to piece together the pattern, he will know who he is and what happened to him. Working with the same handicap as his screenwriter, the character is attempting to construct narrative sense without the luxury of narrative time. Somehow, the reality of his situation must come together for him in a single moment.

CSI, one of the most popular franchises on television, brings this presentist sensibility to the standard crime drama. Where *Law & Order* investigates, identifies, and prosecutes a murderer over a predictable sequence of discoveries, *CSI* uses freeze-frame and computer graphics to render and solve the murder as if it were a puzzle in space. It's not a crime, but a crime *scene*. Potential scenarios—even false ones—are rendered in 3D video maps, as the detectives attempt to deconstruct a single sustained moment.

The TV hits *Lost* and *Heroes* take on this same quality. In *Lost*, characters find themselves on an island where the rules of linear time no longer apply. Successive seasons of the series bring progressively more convoluted permutations on time travel and fate. Solving the mystery of the island and their relationship to it is not the result of a journey through evidence but a "making sense" of the world in the moment. *Heroes* moves back and forth through time in a similar fashion, replacing linear storytelling with the immediacy of puzzle solving. While the various superheroes are indeed preparing to prevent an apocalyptic explosion from destroying New York City, the dramatic action is much more concerned with piecing together a coherent temporal map of the universe in which they are living. The shows are less about what will happen next, or how the story will end, than about figuring out what is actually going on right now—and enjoying the world of the fiction, itself.

True, there may have always been forms of storytelling less concerned with climax than they are with their own perpetuation. The picaresque adventures of Don Quixote gave way to the serialized adventures of Dickens's novels and eventually found new life in American soap operas. But the beauty and reassurance of these sorts of entertainments was that there was always a tomorrow. Somehow, the characters would continue on and never gain so much insight as to become wise. They were often children, or perpetually deluded, or just plain simple. As time sped up and narrative fell apart, however, the soap opera form declined from a height of nineteen different shows in 1970 to just four today.

Taking their place are the soap-opera-like series of pay television, such as the acclaimed *The Wire* and *The Sopranos. The Wire*, which follows drug dealers, corrupt union bosses, and politicians through Baltimore, never doles out justice. A world in which no good deed goes unpunished, *The Wire* is as existential as TV gets—a static world that can't be altered by any hero or any plot point. It just *is*. (The characters may as well be on the series *Oz*, which takes place in the limbo of prison.) Characters experience their reality in terms of their relationship to the "game"—a way of life that is experienced more like a person playing an arcade shooter than going on an epic quest. Likewise, *The Sopranos* was a soap opera about survival in the midst of internecine battles. The characters long to be characters in *The Godfather* movies, who lived by a strict code of ethics and whose careers had more predictable, traditional arcs. The celebrated, controversial last episode of the series was one of TV's most explicit depictions of present shock: in a seemingly innocuous scene, the screen suddenly just goes black. Tony Soprano's existence could end at any moment, without his even being aware that it has ended. No drama, no insight. So, too, for the member of a society without narrative context—at least until he develops alternatives to the linear story.

Still other television creators have taken their cue instead from the epic narratives of Japanese manga comics, developing stories with multiple threads that take years to unfold. Individual episodes of *The X Files* (1993), *Babylon Five* (1994), *Battlestar Galactica* (2004), *Mad Men* (2007), or *Breaking Bad* (2008) may not be capable of conveying a neatly arced storyline, but the slowly moving "meta narrative" creates sustained tension—with little expectation of final resolution.

New incarnations of this approach, such as HBO's sprawling *Game of Thrones* (2011), use structures and tropes more common to player-derived fantasy role-playing games than television. The opening titles sequence of the show betrays this emphasis: the camera pans over an animated map of the entire world of the saga, showing the various divisions and clans within the empire. It is drawn in the style of a fantasy role-playing map used by participants as the game board for their battles and intrigues. And like a fantasy role-playing game, the show is not about creating satisfying resolutions, but rather about keeping the adventure alive and as many threads going as possible. There is plot—there are many plots—but there is no overarching story, no end. There are so many plots, in fact, that an ending tying everything up seems inconceivable, even beside the point.

This is no longer considered bad writing. In fact, presentist literature might even be considered a new genre in which writers are more concerned with the worlds they create than with the characters living within them. As Zadie Smith, author of *White Teeth*, explained in an interview, it is no longer the writer's job to "tell us how somebody felt about something, it is to tell us how the world works."[9] Like other contemporary authors, such as Don DeLillo, Jonathan Lethem, and David Foster Wallace, Smith is less concerned with character arcs than with what she calls "problem solving." Just like the worlds of television's *Lost* or *Heroes*, the worlds of DeLillo's *White Noise* and Lethem's *Chronic City* are like giant operating systems whose codes and intentions are unknown to the people living inside them. Characters must learn how their universes work. Narrativity is replaced by something more like putting together a puzzle by making connections and recognizing patterns.

REALITY BYTES

This same impulse lies at the heart of so-called reality TV—the unscripted, low-budget programs that have slowly replaced much of narrative television. Producers of these shows chose to embrace the collapse of linear narrative once they realized this meant that they were also relieved of the obligation to pay writers to tell a story or actors to perform it. Instead, the purveyors of reality TV simply roll camera in situations and locations that are most likely to generate drama or at least some conflict.

The first of these shows documented something close to reality. *Cops*, created by John Langley back in 1989 and still in production in dozens of countries, follows real police officers as they go on their patrols, chase drug addicts or drunk spouse abusers, and make arrests. The show's opening song lyric, "What you gonna do when they come for you?" reveals the presentist premise: this is instructional video for how to act when you get arrested. The self-satisfied closing monologues of the police assuring us that justice has simply been served replace the character-driven insights and clever reversals of traditional crime drama. The arresting of lowlifes is not a special event but what might be considered a "steady state"—a constant hum or a condition of the presentist universe.

Another reality-TV archetype was born with MTV's *The Real World*, first broadcast in 1992 and still on the air, which represents itself as a slice of its viewing demographic's real lives. The show was inspired by the 1970s documentary series *An American Family*, which set out to document the daily reality of a typical family's experience but hit the jackpot when the parents waged an unexpected and spectacular divorce while the son also realized he was gay, all in prime time. In hopes of yielding similarly sensational effects, *The Real World* selects a group of good-looking eighteen- to twenty-five-year-olds and puts them in an apartment together with dozens of cameras running twenty-four hours a day. Any moment is as potentially significant as any other. It's up to the editors to construct something like narrative, after the fact. Of course, the participants are actually competing for attention, and hoping to get noticed and selected for careers on MTV or in a related industry. So they generate drama as best they can by having sex, fighting, engaging in dangerous behavior, or thinking up something provocative to do that hasn't been done by someone else in one of the other twenty-eight seasons of the show. *The Real World* also solves the television advertiser's dilemma in a medium where traditional commercials no longer work: product placement.

After all, in the channel surfers' DVR-enabled media environment, sponsors no longer have the luxury of captive viewers who will sit through commercials. Many of us are watching entire season's worth of episodes in a single weekend through streaming services such as Hulu or Netflix. The traditional timeline of television schedules vanishes in an on-demand world, so the sponsor must embed advertising into the very fabric of the programming. Reality TV has proved a better backdrop for this, because the disruptions don't compromise the reality of the situation. When we see a real product in a fictional show, we are drawn out of the fantasy and back into real-world considerations. The viewers of *The Real World*, ironically, have no such expectation of the consistency of reality. They know enough about marketing to accept that the clothes the participants wear might be sponsored by fashion companies in the same way that professional sports uniforms are sponsored by Nike. That's the real world, after all. If Donald Trump's "apprentices" are all working to brand a new hamburger, the audience understands that Burger King has paid for the exposure. In presentist TV, programmers lose the ability to distinguish between the program and the commercial—but they also lose the need to do so.

The bigger challenge is creating content compelling enough to watch, and to do so without any setup at all. Without the traditional narrative arc at their disposal, producers of reality TV must generate pathos directly, in the moment. This accounts for the downward spiral in television programming toward the kind of pain, humiliation, and personal tragedy that creates the most immediate sensation for the viewer. What images and ideas can stop the channel surfer in his tracks? The extent of the horror on screen is limited only by the audience's capacity to tolerate the shame of its own complicity. We readily accept the humiliation of a contestant at an *American Idol* audition—such as William Hung, a Chinese American boy who revived the song "She Bangs"—even when he doesn't know we are laughing at him. The more of this kind of media we enjoy, the more spectacularly cruel it must be to excite our attention, and the better we get at evading the moral implications of watching the spectacle.

It calls to mind the questions posed back in the 1960s by Yale psychologist Stanley Milgram, who had been fascinated by the impact of spectacle and authority on soldiers' obedience in Nazi Germany. Milgram wanted to know if German war criminals could have been following orders, as they claimed, and not truly complicit in the death camp atrocities. At the very least, he was hoping to discover that Americans would not respond the same way under similar circumstances. He set up the now infamous experiment in which each subject was told by men in white lab coats to deliver increasingly intense electric shocks to a victim who screamed in pain, complained of a heart condition, and begged for the experiment to be halted. More than half of the subjects carried out the orders anyway, slowly increasing the electric shocks to what they believed were lethal voltages. On a structural level—and maybe also an emotional one—reality TV mirrors much of this same dynamic. No, we're not literally shocking people, but we are enjoying the humiliation and degradation of the participants, from the safe distance of an electronic medium. The question is not how much deadly voltage can we apply, but how shamefully low can we go? Besides, the producers bear the real brunt of responsibility—just as the men in lab coats did in the research experiments.*

And while these sorts of studies were declared unethical by the American Psychological Association in 1973, reality TV does seem to have picked up the thread—less for research purposes than as a last-ditch attempt to generate spectacle without narrative. Because they can't write scripts, reality-show producers must front-load the probability for drama into the very premises of their shows. In this sense, programs like *Big Brother, Survivor,* or *Wife Swap* are as purposefully constructed as psychology experiments: they are setups with clear hypotheses, designed to maximize the probability of drama, conflict, and embarrassment. Let's get women to fawn over a millionaire, perform sex acts with him on television, and then reveal that he's really a construction worker. Let's confine a dozen celebrity addicts in a recovery facility together, make them go off drugs cold turkey, and see what happens! (Yes, these are real shows.)

There is certainly a freedom associated with the collapse of narrative, but it is very easily surrendered to the basest forms of spectacle and abuse. Why bother making a television show at all when kids are more likely to watch a one-minute *Jackass* clip on YouTube of a young man being subjected to the "fart mask"? (Don't ask.) We find nearly every corner of popular culture balancing the welcome release from traditional storylines against the pressure to produce similarly heightened states without them. The emergence of interactivity and deconstruction of the late 1990s led to more than one reaction from programmers and audiences. It resulted in both

---

* Ironically, after I whimsically suggested this connection during a lecture tour in Europe, a French television producer tried it for real in a program called *The Game of Death.* Of course, in this case, the real victims were the people who believed they were the torturers. Under the approving watch of the producers, many contestants delivered what they believed to be lethal doses of electricity to hired actors.

the self-conscious, existentially concerned presentism of *The Sopranos* and *Lost*, as well as the crass, spectacular present shock of *The Jerry Springer Show*, *Faces of Death* videos, *Mob Wives*, *Toddlers and Tiaras*, and even the Paris Hilton sex tapes.

Concerned mythologists and anthropologists foresaw this moment of discontinuity and called upon storytellers to create a new story for this new society—or else. Joseph Campbell believed the first images of the Earth from space utterly shattered our individual cultural narratives and required humanity to develop a universal story about Gaia, the Earth Mother. That clearly hasn't happened. Robert Bly sees manhood as the principal victim of the end of storytelling, as men no longer have a way to learn about the role of the father or the qualities of good leadership. By retelling lost myths, Bly hopes, men can reestablish their connection to these traditions.

But stories cannot truly come to the rescue of people who no longer have the time or trust required to respond to narrativity. What if stories themselves are incompatible with a presentist culture? How then do we maintain a sense of purpose and meaning? Moreover, how do we deal with the trauma of having lost these stories in the first place?

Some of the initial responses to living in a world without narrative are as refreshing as others are depressing. What appears at first like a positive, evolutionary leap in our ability to contend with newfound complexity quickly, even invisibly, often seems to degenerate into exploitation and cynicism. It's not an either/or; in most cases, both are true.

Young people have proved the most adaptable in this regard, maybe because they are less likely to mourn what they never knew about to begin with. It's not just their media consumption that's changing, but their social and physical activities as well. For example, freestyle, independent sports such as skateboarding and snowboarding have surged right along with nonnarrative media. Camcorders, indie media, and YouTube no doubt had something to do with this, as kids had an easy way to record and share their best stunts for all the world to see. A ten-second video shown on a one-inch-high screen can't really do justice to a Major League Baseball double play (you can't even see the ball), but it's the perfect format for a single daring skateboard flip. The individualism of personal media seemed perfectly wed to the individualism of extreme sports. But the shift in sports activity goes deeper than this.

Traditional team sports are not only dependent on mainstream, larger screen television for their success, but also on the top-down style of old-school media and narrative consistency of an enduring and inviolable value system. Gridiron football has distinct battle lines, clear adversaries, regional loyalties, and a winner-take-all ethos. It is a military simulation that follows the same arc as any traditional teaching story—and often with a similar social or commercial purpose. A band of brothers comes together under the guidance of a fatherly coach, who leads them in a locker room prayer before inspiring them with a pep talk worthy of Henry V on Saint Crispin's Day. These tropes and values don't resonate in a postnarrative world. Besides, when we learn that the actual pregame talk in the locker room involves coaches offering a cash bounty to players who can injure the opposing quarterback,[10] the integrity of that story is undermined—no matter how vigorously the league later fines the offenders.

Baseball, meanwhile, found its power in hometown loyalty and American spirit. This was America's pastime, after all. But as the generic power of cash overtakes the sport, teams end up moving from city to city, from historic stadiums to ones named after corporations, while players—free agents—follow the money. Players today rarely come from the cities where they play, making traditional hometown-hero narratives impossible to sustain.

Familiar team rivalries give way to personal careerism, as players compete for record-breaking stats that will win them lucrative product endorsements. Players don't take steroids on behalf of the team but to accumulate home run totals worthy of admittance to the Hall of Fame. The congressional hearings on steroids in baseball revealed the players' own disillusionment with the sport. "Let's face it," José Canseco explained, "when people come to the ballpark, or watch us on TV, they want to be entertained." Jason Giambi echoed the sentiment: "We're in the entertainment business." Disgraced home run legend Barry Bonds sadly admitted, "The last time I played baseball was in college."[11]

In a piece on Boston's 2004 World Series victory, ESPN sportswriter Bill Simmons best voiced the sense of interruption and narrative collapse: "I always thought that, for the rest of my life, I would look at that banner and think only good thoughts. Now, there's a mental asterisk that won't go away. I wish I could take a pill to shake it from my brain."[12] There's a real asterisk, too, in the record books besides the winning seasons of steroid-using teams—as if to extract those years from the timeline of history. Senator John McCain sadly concluded that Major League Baseball was "becoming a fraud in the eyes of the American people."[13] And, of course, every scandal and betrayal is chronicled in painstaking clarity by an always-on cadre of amateur journalists who—unlike the television networks—have nothing to lose by dismantling the illusion of traditional team sports.

As a result, both the National Football League and Major League Baseball have been experiencing declines in attendance over the past four years.[14] The only league on the rise is the NBA, where individual performance, celebrity, and slam dunks take precedence over local affiliation or team spirit. Michael Jordan's fade-away jumpers will be remembered longer than his tenuous connection to Chicago. But even there, strident individualism feels at odds with the central premise of team sports. When NBA superstar LeBron James famously used a televised special—*The Decision*—to announce he was leaving his home state of Ohio to play for Miami, the Cleveland Cavaliers fans never forgave him.

Freestyle sports, like skateboarding, snowboarding, rock climbing, and mountain biking, are more compatible with a world in which team loyalty and military victory have given way to self-expression and the thrill of the moment. Team sports take hours to watch and require a very particular sort of commitment to play. A kid must sign up at school, submit to a coach, and stay for the season. By contrast, extreme sports are improvisational in nature, and more about texture, pleasure, and style than about victory over an adversary. They emphasize process, form, and personal achievement, and resist efforts to standardize play. When the world's first great snowboarders were asked to participate in the initial Olympics competition for snowboarding in 1998, they refused, because they were concerned it would redefine a freestyle sport as rules-based and combatively competitive.

Extreme sports also tend to celebrate the very discontinuities—the roughness—that well-crafted narrative arcs try to smooth over. Compared with traditional parallel skiers, who pursue a streamlined descent down the slope, snowboarders actively thrash their course. They seek out the rough, icy patches that regular skiers avoid, in order to test their mettle and improvise new maneuvers. They cherish the unpredictability of the slope the same way that skateboarders thrash the discontinuities of the urban landscape. Every crack in the sidewalk, every hydrant, and every pothole is an opportunity. Boarders' magazines of the mid-1990s contained articles relating skateboarding to chaos math, and snowboarding to Buddhism.

The highly experiential now-ness of extreme sports was equally susceptible to the downward pull of present shock. Professional basketball has a hard enough time maintaining a team ethic in the face of showboating and hotdogging antics on the court and off. Only heavy fines are enough to keep players thinking about team priorities, like actually winning games. Without any such disincentives, extreme sports stars soon disregarded the anticommercial ethos of their predecessors and began accepting lucrative sponsorships from clothing companies and appearing on television shows dedicated to chronicling their exploits.

Meanwhile, the net celebrates kids whose antics are the most sensationalist and, as a result, often reckless and self-destructive. An entire genre of YouTube video known as Epic Fail features amateur footage of wipeouts and other, well, epic failures. "FAIL Blog," part of *The Daily What* media empire, solicits fail videos from users and features both extreme sports stunts gone awry along with more random humiliations—like the guy who tried to shoplift an electric guitar by shoving it down his pants. Extreme sports clips are competing on the same sensationalist scale and result in popular classics such as "tire off the roof nut shot" and "insane bike crash into sign." Daring quickly overtakes what used to be skill. In "planking" photos and videos, participants seek to stay frozen in a horizontal plank position as they balance on a flagpole, over a cliff, or on top of a sleeping tiger. For "choking" videos, young people strangle one another to the point of collapse and, sometimes, death.[15]

Maintaining a dedication to craft over crassness is not impossible, but pretty difficult when sports enthusiasts find themselves competing for the same eyeballs and accolades as plankers and chokers. The predicament they face casts an unlikely but informative light on the way adults, too, are coping with the demise of the stories we watch and use to participate in the world around us.

REAL-TIME FEED: THE CNN EFFECT

While the shift away from grand narrative in spectator sports may not radically impact the world, the shift away from narrative in spectator democracy just could. What began as a simple expansion of the way we were allowed to consume television news has resulted in profound changes to the ways in which both policy and politics are conducted, or, sadly in most cases, not conducted at all. A presentist mediascape may prevent the construction of false and misleading narratives by elites who mean us no good, but it also tends to leave everyone looking for direction and responding or overresponding to every bump in the road.

Until recently, of course, television news tended to reinforce traditional narrative values. Like the newsreels once shown in movie theaters, broadcast news compiled and contextualized footage from the field. By the early 1960s, the three main networks—CBS, NBC, and ABC—each had its own fifteen- to thirty-minute news program every weekday evening, anchored by a reassuring middle-aged man. These broadcasts enjoyed such authority that Walter Cronkite could end his broadcast, nonironically, by saying, "And that's the way it is." The daily news cycle gave everyone, from editors to politicians, the opportunity to spin and contextualize news into stories. This is what journalism schools taught: how to shape otherwise meaningless news into narratives.

Morning newspapers would have even more time to digest, format, and editorialize on the news of the preceding day, so that the public wouldn't simply be confronted with the globe's many catastrophes. We would be told what was being done about them or how they ended up, and those in charge were located and given a chance to reassure us. News editors also chose when to hold back a story altogether, for fear that its unresolved nature might worry us too greatly. Foreign dictators were not granted US airtime, for example, and scandals about politicians were held indefinitely or forever. Just as the *New York Times* promised us news that was "fit to print," television news shows sought to promote the interests, welfare, and contentment of America by creating a coherent narrative through which we could understand and, hopefully, dismiss the news before going to bed. Thank you and good night.

None of this was necessarily intended to be devious. The widespread belief among both the political and the media industries was that the public was not sophisticated enough to grasp the real issues of the day. Walter Lippman, America's first true public relations specialist, had convincingly argued in his 1922 book *Public Opinion*, "The real environment is altogether too big, too complex, and too fleeting for direct acquaintance."[16] Incapable of grasping news directly, the public was to be informed about issues only after a benevolent elite had crafted all this information and its implications into simple and palatable stories. In this view, the people are incapable of participating as informed members of a democracy, and their votes should not be left up to their discretion. Instead, public relations specialists would be hired to get people to vote in their own best interests. So, for example, after winning the presidency on a peace platform, Woodrow Wilson subsequently decided to go to war. He hired Lippman and his protégé, Edward Bernays, to manufacture public consent for American participation in World War I.

Later, coverage of the Vietnam War threatened this arrangement, as daily footage of American soldiers enduring and even perpetrating war atrocities proved too much for the evening news to contextualize. There was still a story being told, but the story was out of the government's—or its propagandists'—control. If anything, the story that the television news was telling ended up more accurate than the one President Johnson's staff was feeding him. The cognitive dissonance between the stories we were trying to tell ourselves about who we were as a nation and a people began conflicting with the stories that we were watching on TV. In a world still organized by stories, news about Vietnam atrocities and Watergate crimes can only mean there are bad people who need to be punished.

This cognitive dissonance amounted to a mass adolescence for America: the stories we were being told about who we were and what we stood for had turned out to be largely untrue. And like any adolescent, we felt ready to go out and see the world for ourselves. What a perfect moment for Ted Turner to arrive with an unfiltered twenty-four-hour news channel. Like every other escape from media captivity, the launch and spread of CNN in the 1980s appeared to offer us liberation from the imposed narratives of our keepers. Instead of getting the news of the day neatly packaged by corporate networks into stories with tidy endings, we would now get live feeds of whatever was happening of importance, anywhere in the world. Although CNN would lack the network news's budgets, editorial experience, and recognizable anchormen, it would attempt to parlay its unique position to its advantage. CNN was not under traditional corporate control, so it could present news without worrying about who or what was impacted; its always-on format meant the newsroom was not under the obligation to craft events into satisfying packages for a single evening broadcast; its position on the cable dial lowered expectations for high-budget production values; and, finally, its freedom from traditional narrative made it less suspect in the post-Watergate era. Like Ted Turner himself, CNN would be renegade, free of corporate or government control, and utterly uncensored.

This is why the network came under such widespread attack in its earlier days from traditional news media, academics, and politicians. Contrary to journalistic standards of the day, CNN let Saddam Hussein speak directly to the American people. During Operation Desert Storm in 1991, CNN was the only network to broadcast the first bombardments of Baghdad from a hotel window, having made arrangements for transmission access with the Iraqi government that were later criticized. CNN also carried immediate and around-the-clock coverage of the Battle of Mogadishu, the protests at Tiananmen Square, and the fall of communism in Eastern Europe. This saturation with live, uncensored, and unconsidered images from around the world impacted public opinion profoundly and actually forced government leaders to make decisions more quickly. Officials at the Pentagon eventually dubbed this phenomenon "the CNN effect," as then secretary of state James Baker explained, "The one thing it does, is to drive policymakers to have a policy position. I would have to articulate it very quickly. You are in real-time mode. You don't have time to reflect."[17] Baker isn't simply talking about needing to work and think faster; he's expressing the need to behave in real time, without reflection. Policy, as such, is no longer measured against a larger plan or narrative; it is simply a response to changing circumstances on the ground, or on the tube. Of course, the Internet, Facebook status updates, and Twitter feeds amplify this effect, bringing pings and alerts from around the world to people's desktops and smart phones without even the need for a CNN truck or a satellite feed. Just as CNN once forced network news broadcasters to carry images they might have otherwise held back, now YouTube forces the cable networks to show amateur clips, even if they do so with disclaimers.

The focus on immediate response engendered by always-on news becomes the new approach to governance. Pollsters such as Republican operative Frank Luntz[18] take real-time, moment-to-moment measurements of television viewers' responses as they watch news debates. Holding small devices with dials called "people meters," sample audience members register their immediate impressions of candidates, coverage, and calamity. Policy makers then use this information to craft their responses to crises as they unfold. No one has time to think, and insisting on a few hours or even an entire day to make a decision is regarded as a sign of indecision and weakness. The mythic emergency phone call to the president at 2 a.m. must not only be answered, but also responded to immediately, as if by instinct.

As a result, what used to be called statecraft devolves into a constant struggle with crisis management. Leaders cannot get on top of issues, much less ahead of them, as they instead seek merely to respond to the emerging chaos in a way that makes them look authoritative. While grand narratives may have prompted ethnocentric and jingoistic attitudes from ideological policy makers (neoconservatism

being just one of the more recent varieties of world writing), the
lack of any narrative at all subjects them to the constant onslaught of
random disasters. The effort to decisively end a story is futile. George
W. Bush posed his control over narrativity just three days after 9/11
when he stood at Washington National Cathedral and told America,
"This conflict was begun on the timing and terms of others; it will
end in a way and at an hour of our choosing."[19] He later landed on an
aircraft carrier and stood in front of a Mission Accomplished banner,
as if to punctuate the conclusion of this story, but reality would not
submit. In a presentist world, it is impossible to get in front of the
story, much less craft it from above.

Likewise, without long-term goals expressed for us as readily
accessible stories, people lose the ability to respond to anything but
terror. If we have no destination toward which we are progressing,
then the only thing that motivates our movement is to get *away* from
something threatening. We move from problem to problem, avoiding
calamity as best we can, our worldview increasingly characterized by
a sense of panic. Our news networks and Internet feeds compound
the sense of crisis by amplifying only the most sensational and negative events, which garner the highest ratings and click-throughs,
generating still more of the same. Yes, the news has always been
dominated by darkness and disaster; newspapers with a dead body on
the cover page sell better than those announcing a successful flower
show. But now the feedback on viewer ratings is instantaneous, and
its relationship to ad revenues is paramount, as once-independent
news channels like CNN become mere subsidiaries of NYSE con-
glomerates such as Time Warner. The indirect value that a quality
news program may have to the reputation of an entire television fran-
chise ceases to have meaning—or at least enough meaning to compensate for poor ratings in the short term. Blatant shock is the only
surefire strategy for gaining viewers in the now. In addition, the 24/7
news cycle creates the sense of a constant stream of crises that are

inescapable, no matter where we go. I remember a time in my youth when taking a vacation meant losing touch with whatever was going on. We would return home to mail, phone messages, and the major news stories. Now, as kids on line at Disneyworld busy texting their friends back home seem to attest, there's almost no way or cause to leave all that behind.

The world is now connected by the news feeds of twenty-four-hour networks, and so together we watch the slow-motion, real-time disasters of Hurricane Katrina, Deepwater Horizon, and the Fukushima nuclear plant—as well as what feels like the utter ineffectualness of our leaders to do anything about any of it. CNN put up a live feed of the BP well spewing out its oil into the gulf and kept it in a corner of the broadcast continuously for months. The constancy of such imagery, like the seemingly chronic footage of Katrina victims at the New Orleans Superdome holding up signs begging for help, is both unnerving and desensitizing at the same time. With each minute that goes by with no relief in sight our impatience is stoked further and our perception of our authorities' impotence is magnified.

Talk radio and cable channels such as Fox News make good business out of giving voice to presentist rage. Opinionated, even indignant, newsreaders keep our collective cortisol (stress hormone) levels high enough to maintain a constant fight-or-flight urgency. Viewers too bored or impatient for news reporting and analysis tune in to evening debate shows and watch pundits attack one another. The pugilism creates the illusion of drama, except the conflict has no beginning or end—no true origin in real-world issues or legitimate effort at consensus. It's simply the adaptation of well-trodden and quite obsolete Right-Left debate to the panic of a society in present shock. What used to be the Left argues for progress: MSNBC's brand motto encourages us to "lean forward" into the future. What used to be the Right now argues primarily for the revival of early-twentieth-century values, or social conservatism. Whether looking

back or looking ahead, both sides promise relief from the shock of
the present.

OCCUPY REALITY

The problem with leaving the present altogether, however, is that it
disconnects us from reality. For all the reality shows, twenty-four-
hour news channels, issues-related programming, and supposed information overload online, there's precious little for people to actually
rely on or use effectively. Are real estate prices going up or down?
Who is winning in Afghanistan? Do Mexicans take American jobs?
It all depends on who is talking, as the descent of what used to be
professional journalism into professional opining generates the sense
that there is no objective truth. Every day, thanks to their immersion
in this mediated distortion field, fewer Americans agree that the environment needs to be protected or that biological species evolve.
From 1985 to 2005, the number of Americans unsure about evolution increased from 7 percent to 21 percent,[20] while those question-
ing global warming increased from 31 percent in 1997 to 48 percent
in 2010.[21] These impressions are formed on the basis of religious programming posing as news reporting and cable-channel debates about
email scandals, while back in the real world, aquifers are disappear-
ing and first-line antibiotics are becoming ineffective against rapidly
mutating bacteria. In the relativistic haze of participatory media, it's
all just a matter of opinion. You are entitled to yours and I am entitled to mine. This is a democracy, after all. As even the jaded, former
public relations giant Richard Edelman now admits, "In this era of
exploding media technologies there is no truth except the truth you
create for yourself."[22]

The Internet welcomes everyone into the conversation. An op-ed
in the *New York Times* may as well be a column on the *Huffington*

Post, which may as well be a personal blog or Twitter stream. Everyone's opinion may as well matter as much as everyone else's, resulting in a population who believes its uninformed opinions are as valid as those of experts who have actually studied a particular problem. (I can even sense readers bridling at the word "experts" in the preceding sentence, as if I have fallen into the trap of valuing an elite over the more reliable and incorruptible gut sense of real people.) College students often ask me why anyone should pay for professional journalism when there are plenty of people out there, like themselves, willing to write blogs for free? One answer is that government and corporations are investing millions of dollars into their professional communications campaigns. We deserve at least a few professionals working fulltime to evaluate all this messaging and doing so with some level of expertise in ascertaining the truth.

Young people are not alone in their skepticism about the value of professional journalism. A 2010 Gallup Poll showed Americans at an under 25 percent confidence in newspapers and television news—a record low.[23] Pew Research shows faith in traditional news media spiking downward as Internet use spikes upward, and that a full 42 percent believe that news organizations hurt democracy. This is twice the percentage who believed that in the mid-1980s, before the proliferation of the net.[24]

As cultural philosopher Jürgen Habermas offered during his acceptance speech of a humanitarian award in 2006, "The price we pay for the growth in egalitarianism offered by the Internet is the decentralized access to unedited stories. In this medium, contributions by intellectuals lose their power to create a focus."[25] To be sure, the rise of citizen journalism brings us information that the mainstream media lacks either the budget for or fortitude to cover. Initial reports of damage during Hurricane Katrina came from bloggers and amateur videographers. However, these reports also inflated body counts and spread rumors about rape and violence in the Superdome that were later revealed not to have occurred.[26] Footage and reporting from the Arab Spring and the Syrian revolution—where news agencies were limited or banned—were almost entirely dependent on amateur journalists. But newsgathering during a bloody rebellion against a violently censorious regime is an outlier example and hardly the basis for judging the efficacy of amateur journalism in clarifying issues or explaining policy.

If anything, such heroism under fire, combined with the general public's access to blogging technology and professional-looking website templates, gives us all the false sense that we are capable of researching and writing professional-quality journalism about anything. In fact, most of us are simply making comments about the columns written by other bloggers, who are commenting on still others. Just because we all have access to blogging software doesn't mean we should all be blogging, or that everyone's output is as relevant as everyone else's. Today's most vocal critic of this trend, *The Cult of the Amateur* author Andrew Keen, explains, "According to a June 2006 study by the Pew Internet and American Life Project, 34 percent of the 12 million bloggers in America consider their online 'work' to be a form of journalism. That adds up to millions of unskilled, untrained, unpaid, unknown 'journalists'—a thousandfold growth between 1996 and 2006—spewing their (mis)information out in the cyberworld." More sanguine voices, such as City University of New York journalism professor and BuzzFeed blogger Jeff Jarvis, argue that the market—amplified by search results and recommendation engines—will eventually allow the better journalism to rise to the top of the pile. But even market mechanisms may have a hard time functioning as we consumers of all this media lose our ability to distinguish between facts, informed opinions, and wild assertions.

Our impatient disgust with politics as usual combined with our newfound faith in our own gut sensibilities drives us to take matters into our own hands—in journalism and beyond. In a political world where ideological goals are replaced by terror and rage, it's no wonder the first true political movement to emerge out of present shock would be the Tea Party. This is the politics of PTSD, inspired by a no-nonsense brand of libertarianism espoused by Texas congressman Ron Paul. Taking its name from the Boston Tea Party of 1773, when American colonists dumped British tea into the harbor in a tax revolt, today's Tea Party movement shares the antiauthoritarian impulse of its namesake and then expresses it as a distrust of government in all forms. While the Tea Party may have originated as an antitax movement, it has been characterized over time more by a disdain for consensus and an almost deliberate effort to remain ignorant of facts that may contradict its oversimplified goals.

Tea Partiers, such as Michele Bachmann, either misunderstood or intentionally misrepresented the concept of a debt ceiling as a vote to authorize additional spending (when it is actually a vote to pay what has already been spent). The solution to the seemingly perpetual debt crisis? Shut down government. Healthcare system too complicated? End it. (Except, of course, for Medicare, which doesn't count.) Russia and China are evil, Arabs are scary, Mexicans are taking Americans' jobs, and climate change is a hoax. As Columbia University historian Mark Lilla has chronicled, the combination of amplified self-confidence and fear of elites is a dangerous one. In his view, the Tea Partiers "have two classic American traits that have grown much more pronounced in recent decades: blanket distrust of institutions and an astonishing—and unwarranted—confidence in the self. They are apocalyptic pessimists about public life and childlike optimists swaddled in self-esteem when it comes to their own powers." [27]

If the Tea Party is to be disparaged for anything, it is not for being too conservative, too right wing, or too libertarian, but simply too immature, quick-triggered, and impatient for final answers. Traumatized by the collapse of the narratives that used to organize reality and armed with what appears to be access to direct democracy, its members ache for harsh, quick fixes to age-old problems—something they can really *feel*—as if fomenting a painful apocalypse would be better than enduring the numbing present.

More intellectually grounded conservatives and GOP regulars fear the Tea Party more than they fear Democrats, for they understand that this knee-jerk race to results undermines the very foundation and justification for representative democracy. As former George W. Bush speechwriter David Frum laments:

> A political movement that never took governing seriously was exploited by a succession of political entrepreneurs uninterested in governing—but all too interested in merchandising. Much as viewers tune in to *American Idol* to laugh at the inept, borderline dysfunctional early auditions, these tea-party champions provide a ghoulish type of news entertainment each time they reveal that they know nothing about public affairs and have never attempted to learn. But Cain's gaffe on Libya or Perry's brain freeze on the Department of Energy are not only indicators of bad leadership. They are indicators of a crisis of followership. The tea party never demanded knowledge or concern for governance, and so of course it never got them.[28]

Representative democracy has a hard enough time justifying itself in a digitally connected world where representation no longer means sending someone on a three-day carriage ride to the capital. Having cynically embraced the Tea Party as a means to an end, Republicans now face erosion of party integrity from within. Meanwhile, as if aware of the role that the twenty-four-hour news cycle played in having generated this phenomenon, CNN partners with the Tea Party to arrange televised presidential debates. For the one thing the Tea Party appears to want more than the destruction of government is to elect Tea Party members to positions within it.

The impatient rush to judgment of the Tea Party movement is only as unnerving as the perpetually patient deliberation of its counterpart present shock movement, Occupy Wall Street. Opposite reactions to collapse of political narrative, the Tea Party yearns for finality while the Occupy movement attempts to sustain indeterminacy.

Inspired by the social-media-influenced revolutions of the Arab Spring, Occupy Wall Street began as a one-day campaign to call attention to the inequities inherent in a bank-run, quarterly-focused, debt-driven economy. It morphed into something of a permanent revolution, however, dedicated to producing new models of political and economic activity by its very example. Tea Partiers mean to wipe out the chaotic confusion of a world without definitive stories; the Occupiers mean to embed themselves within it so that new forms may emerge. It's not an easy sell. The Tea Party's high-profile candidates and caustic rhetoric are as perfectly matched for the quick-cut and argument-driven programming of the cable news networks as the Occupiers are incompatible. Though both movements are reactions to the collapse of compelling and believable narratives, the Tea Party has succumbed to and even embraced the crisis mentality, while Occupy Wall Street attempts to transcend it.

This is at least part of why mainstream television news reporters appeared so determined to cast Occupy Wall Street as the random, silly blather of an ungrateful and lazy generation of weirdos. As if defending against the coming obsolescence of their own truncated news formats, television journalists reported that the movement's inability to articulate its agenda in ten seconds or less meant there was no agenda at all. In a segment titled "Seriously?!" CNN business anchor Erin Burnett ridiculed the goings-on at Zuccotti Park. "What are they protesting?" she asked. "Nobody seems to know." Like *The Tonight Show* host Jay Leno testing random mall patrons on American history, Burnett's main objective was to prove that the protesters didn't know that the US government had been reimbursed for the bank bailouts. More predictably, perhaps, a Fox News reporter appeared flummoxed when the Occupier he interviewed refused to explain how he wanted the protests to end. Attempting to transcend the standard political narrative, the protester explained, "As far as seeing it end, I wouldn't like to see it end. I would like to see the conversation continue."[29]

In this sense, regardless of whether its economic agenda is grounded in reality, Occupy Wall Street does constitute the first truly postnarrative political movement. Unlike the civil rights protests, labor marches, or even the Obama campaign, it does not take its cue from a charismatic leader, it does not express itself in bumper-sticker-length goals, nor does it understand itself as having a particular endpoint. The Occupiers' lack of a specific goal makes it hard for them to maintain focus and cohesion. The movement may be attempting to embrace too wide an array of complaints, demands, and goals: the collapsing environment, labor standards, housing policy, government corruption, World Bank lending practices, unemployment, increasing wealth disparity, and so on. But these many issues are connected: different people have been affected by different aspects of the same system—and they believe they are all experiencing symptoms of the same core problem. But for journalists or politicians to pretend they have no idea what the movement wants is disingenuous and really just another form of present shock. What upsets banking's defenders and traditional Democrats alike is the refusal of this movement to state its terms or set its goals in the traditional language of campaigns.

That's because, unlike a political campaign designed to get some person in office and then close up shop (as in the election of Obama and subsequent youth disillusionment), this is not a movement with a traditional narrative arc. It is not about winning some debate point and then going home. Rather, as the product of the decentralized networked-era culture, it is less about victory than sustainability. It is not about one-pointedness, but inclusion. It is not about scoring a victory, but groping toward consensus. It is not like a book; it is like the Internet.

Occupy Wall Street is not a movement that wins and ends; it is meant more as a way of life that spreads through contagion and creates as many questions as it answers. The urban survival camps they set up around the world were a bit more like showpieces, congresses, and beta tests of new ideas or revivals of old ones. Unlike a traditional protest, which identifies the enemy and fights for a particular solution, Occupy Wall Street just sits there talking with itself, debating its own worth, recognizing its internal inconsistencies, and then continuing on as if this were some sort of new normal. It is both inspiring and aggravating.

Occupy's General Assembly methodology, for example, is a highly flexible approach to group discussion and consensus building borrowed from the ancient Greeks. Unlike parliamentary rules that promote debate, difference, and decision, the General Assembly forges consensus by stacking ideas and objections as they arise, and then making sure they are all eventually heard. The whole thing is orchestrated through simple hand gestures. Elements in the stack are prioritized, and everyone gets a chance to speak. Even after votes, exceptions and objections are incorporated as amendments.

On the one hand, the process seems like an evolutionary leap forward in consensus building. Dispensing with preconceived narratives about generating policy demands or settling the score between Right versus Left, this process eschews debate (or what Enlightenment philosophers called "dialectic") for consensus. It is a blatant rejection of the binary, winner-takes-all, political operating system that has been characterizing political discourse since at least the French National Assembly of the 1700s. But it is also a painstakingly slow, almost interminably boring process, in which the problem of how to deal with noise from bongo drummers ends up getting equal time with how to address student debt. It works well for those who are committed to sitting in a park doing little else with their days and nights, but is excruciating for those committed to producing results. Engaged with this way, the present lasts a whole long time.

The ambiance and approach of the Occupiers is more like a university—one of life's great pauses—than a political movement. Both online and offline spaces consist largely of teach-ins about the issues they are concerned with. Young people teach one another or invite guests to lecture them about subjects such as how the economy works, the disconnection of investment banking from the economy of goods and services, possible responses to mass foreclosure, the history of centralized interest-bearing currency, and even best practices for civil disobedience.

The approach is unwieldy and unpredictable but oddly consistent with the values of a postnarrative landscape. The Occupy ethos concerns replacing the zero-sum, closed-ended game of financial competition with a more sustainable, open-ended game of abundance and mutual aid. In the traditional political narrative, this sounds like communism, but to the Occupiers, it is a realization of the peer-to-peer sensibility of the social net. It is not a game that someone wins, but rather a form of play that—like a massive multiplayer online game—is successful the more people get to play, and the longer the game is kept going.

INFINITE GAMES

Computer games may, in fact, be popular culture's first satisfactory answer to the collapse of narrative. Believe what we may about their role in destroying everything from attention spans and eyesight to social interactions and interest in reading, video games do come to the rescue of a society for whom books, TV, and movies no longer function as well as they used to. This is not simply because they are brighter and louder; the sounds and imagery on kids' TV these days have higher resolution and are even more densely packed. Video games have surpassed all other forms of entertainment in market share and cultural importance because they engage with players in an open-ended fashion, they communicate through experience instead of telling, and they invite players into the creative process. While video games do occur over linear time, they are not arced like stories between a past and the future. When they are off, they are gone. When they are on, they are in the now.

Although religious historian James Carse came up with the concept of "infinite games" well before computer games had overtaken television, music, and movies as our dominant entertainment industry, his two categories of play help explain why electronic gaming would gain such favor in an era of present shock. Finite games are those with fixed endings—winners and losers. Most every game from tennis to football works this way. Victory is the scarcity: there can be only one winner, so players compete for the win. Infinite games, on the other hand, are more about the play itself. They do not have a knowable beginning or ending, and players attempt to keep the game going simply for the sake of the play. There are no boundaries, and rules can change as the game continues. Carse's point is to promote the open-ended, abundant thinking of infinite games. Instead of competing against one another and aching for the finality of conclusion, we should be playing with one another in order to maximize the fun for all. Instead of yearning for victory and the death of finite games, we should be actively enjoying the present and trying to sustain the playability of the moment. It's an approach that favors improvisation over fixed rules, internal sensibilities over imposed morals, and playfulness over seriousness.

While there is no such thing as a perfectly infinite game (except maybe for life itself), there are many increasingly popular forms of play that point the way to Carse's ideal. For anyone but professional performers, improvisational storytelling usually ends with early childhood and is replaced with books, television, and organized play. But beginning in the mid-1970s (right around the time that TV remotes became standard features), a new form of game emerged called the fantasy role-playing game, or RPG. Inspired by the rules written for people who play war games with medieval miniature figures, Dungeons & Dragons, the first published RPG, was a simple rule set that allowed players to imagine and enact adventures on a tabletop.* Unlike conventional games with sides and rules and winners, Dungeons & Dragons was really just a starting place for interactive storytelling. Less like a performance than one of artist Allan Kaprow's "Happenings," D&D provided an excuse and a context for people to gather and imagine adventures together. Each player began by creating a character sheet that defined his identity and attributes. Some attributes were purely creative (a dwarf with blond hair who wears a red hat), while others determined a character's abilities in the game, such as strength level, magical skills, or intelligence. A Dungeon Master led the proceedings and refereed interactions. Beyond that, characters went on adventures and engaged in conflicts as fanciful as they could imagine.

The popularity of RPGs was matched only by the consternation they generated among parents and educators. Along with heavy metal music, Dungeons & Dragons was blamed for a collapse of ethics among young teens, drug abuse, and even murder. The underlying fear of RPGs, however, probably had a lot more to do with how open-ended they were. Kids went into the basement or game room for hours, became deeply involved in fantasy adventures that were sustained over months or even years, and the boundary between the game and real life seemed to erode as players wore costume pieces from the game to school, or utilized game strategies in daily activities. RPGs did not respect our notions of time boundaries. When was the game over? Who wrote the rules? How does a

---

* Dungeons & Dragons was itself inspired by a set of rules written by Jeff Perren for people who played with miniature medieval figures, which was later expanded and published as the game Chainmail by Gary Gygax.

person win? If a game doesn't teach winning, is it simply creating losers?

RPG players were aficionados of presentism, not merely tolerating but delighting in stories without endings. While the adventures they invented together had little arcs and minivictories for one character or another, they were valued most for their ability to sustain themselves—and everyone's interest—over long periods of time. Dungeon Masters measured their reputations in terms of how long they could keep a game group together. They had one terrific advantage in this regard over regular storytellers: their audience actively participated in the creation of the story. Instead of leading a passive audience through a vicarious adventure and then inserting a particular value into the climax, the Dungeon Master simply facilitates the imaginative play of a group of peers. He sets up the world in which the play takes place. In this sense, it was the purest precursor to the interactive media that was to follow.

RPG players were a natural fit and ready audience for video games, which tend to require a similarly open and participatory approach to story. While not all video games wrestle directly with issues of narrativity, they all must contend with audience members who have the freedom to make their own, differing choices over the course of the game. Choices may be as limited as which asteroids to shoot at or as expansive as how to ally a number of raiding groups to fight in a war. Wherever in the spectrum of free will and interactivity they fall, however, video games—like RPGs—reverse the rules of Aristotelian narrative. A traditional narrative leads inevitably to its ending. That's the whole point: the character makes the best choices possible but meets a fate that seems almost destined, at least in retrospect. The audience must conclude this is the only way things could have gone given the situation and the characters. If the hero makes a wrong choice, it's considered a hole in the plot.

Video games are just the opposite. While the game writer may have an ending or final level he wants everyone to get to at some point, moving through this world is supposed to feel like free will. Each scene opens up a series of choices. Instead of watching a character make the *only* right choice in each scene, the player is the main character, confronted with a myriad of choices. While a traditional story narrows toward the destined ending, the game branches open to new possibilities. When we read a book or watch a movie, the best choice for each character already exists; it just hasn't been revealed. When we play a game, that choice is happening in real time.

The many different types of video games exercise presentism differently. "Shooters," where the player runs around and shoots monsters or other enemies, may seem the most present-tense but actually offer the least amount of player authorship. While the player can kill things on each level in any number of ways, this only brings him to the next predetermined level. "God" games, like SimCity and Civilization, let players build and supervise worlds. The player may be charged with planning a city, managing a civilization from its inception, or even evolving life from the beginning (as in Spore). The biases of these worlds are determined by the choices the player makes. Violent choices yield a violent world; focusing on business may create a world more dominated by economics; and so on.

But the most compelling and still largest sector of gaming is social games. In massively multiplayer online role-playing games, or MMORPGs, like World of Warcraft and Guild Wars, players join groups and set out on missions together. While the world in which they play is created by game companies, and the various monsters and natural phenomena they encounter have been preprogrammed, the interactions between the characters are all up to the players. Players choose whether to go on quests, to make friends, or to form groups that go on to attack others. The games are social in that participants really are playing with other people—even if separated by computers, Internet connections, and thousands of miles of physical distance.

Even the social games played on the Facebook platform would qualify as a crude form of this style of participatory storytelling. FarmVille and Mafia Wars have been huge successes (enough to justify a billion-dollar stock offering by the company that released them). Like a god game, FarmVille lets players create and tend a farm, sell crops, then buy special decorations with the game money. The payoff, such as it is, is that because the game is happening on a social network, friends can see one another's farms. Similarly, Mafia Wars let players fight with and against one another over territory and status. The main way to gain status, and support the game company, is to recruit more players to be in one's mafia gang. Though they offer a whole lot less creative latitude than true MMORPGs, these games still give people who might never have played a video game the chance to experience entertainment that unfolds in the present tense.

These experiences needn't be entirely devoid of meaning and values, either. Just because an experience lacks narrativity doesn't mean it can't communicate and do so powerfully. Serious games, named for their intended effects, are computer games that attempt to convey serious things. Instead of inserting messages into games the way an author might insert a message into a book, games try to communicate through experience. So instead of watching a character get hoisted on his own petard for being too arrogant, the player is to experience this reversal and recognition himself.

There are games about health, violence, ethics, pollution, and pretty much any serious subject you can imagine. Games for Change, for example, is an organization and website that collects and curates games that engage players with contemporary social issues. One of the most famous such games so far, Darfur is Dying, re-creates the experience of the 2.5 million Sudanese refugees. The player is the refugee, whose gender and other preconditions end up having a lot to do with what happens (rape, murder, or escape). In some sense the game is unwinnable, but the interactions and experience may convey more about the refugee plight and the futility of the situation than a hundred hours of news footage. The game, like many others, provides opportunities for players to donate, Tweet, or otherwise use the Internet to participate in the cause.

Games offer a healthier, or at least more active, response to the collapse of narrativity confounding much of the rest of popular culture. They also offer us an inkling of how we may avert present shock altogether and instead adopt approaches that successfully reorient us to the all-at-onceness of life today. Instead of panicking at the death of the story, players *become* the story and delight in acting it out in real time. The people designing the game can still communicate values if they choose to; they simply need to do it by offering choices instead of making them in advance.

This approach is applicable almost anywhere narrative is failing. In the world of politics, this would mean taking the tack of the Occupiers prototyping new modes of activism—eschewing ends-justify-the-means movements and developing a normative behavior, instead. In retail, the equivalent would mean deemphasizing brand mythologies and focusing instead on what is called brand experience—the actual pathway the customer takes through the real or virtual shopping environment. It's not about the story you tell your customer; it is about the experience you give him—the choices, immersion, and sense of autonomy. (It also means accepting transparency as a new given, and social media as the new mass communications medium, as we'll see in chapter 4.) In medicine, it means enlisting patients in their own healing process rather than asking them to do nothing while blindly accepting the magical authority of the doctor and a pharmaceutical industry. Understanding these cultural, political, and market dynamics through the lens of gaming helps us transition from the world of passively accepted narrative to one that invites our ongoing participation.

Games point the way toward new ways of accomplishing what used to be done with stories. They may not be a cure-all, but they can successfully counteract some of the trauma we suffer when our stories come apart. Our disillusionment is offset by a new sense of participation and self-direction.

In fact, gaming's promise as an antidote to post-traumatic stress on a cultural scale finds support in its increasing use as an actual treatment for post-traumatic stress disorder on an individual level. At a University of Southern California lab called the Institute for Creative Technologies, a psychologist named Albert "Skip" Rizzo has been using virtual-reality simulations to treat Iraq War veterans suffering from severe PTSD. He started with the immersive video war game Full Spectrum Warrior, and then adapted it for use in therapy sessions. The patient wears 3D virtual-reality goggles and describes the scene of his trauma as the therapist builds that world around him. The patient holds a game controller and walks or drives or shoots as he did in the war zone. The therapist uses the game to re-create the entire scene, allowing the patient to relive all of the horror of the lost and tragic moment in the safety and from the distance of a computer simulation.

I had the chance to use the gear myself while working on a documentary about how digital technologies change the way we live. Rizzo operated the simulation as I described a car accident that had occurred many years ago, in which I lost my best friend. I was originally planning on criticizing the technology for getting in the way of the human contact between therapists and their patients—but I was wrong. I had told the story of my car accident to many people, even a few psychotherapists, but never felt anything about the incident had been resolved. Just telling the story somehow was not enough. But in the simulator, I was able to tell Dr. Rizzo that the sky was a bit darker—*it wasn't quite dawn*. He darkened the sky. *Pinker*, I said. He made it pinker. *And there were some thinned-out shrubs on the side of the road*. He added them. *And my friend had a paler complexion*. Done. Rizzo used another device to generate the smell of the desert and the juniper bushes I described. It was as if I were there again.

More important, Rizzo was there with me the whole time. He

wasn't just making the simulation for me; he was in the simulation *with* me. The human connection was actually more profound than when I had told the story to friends and even a therapist before, because I knew for sure that he could *see* what I *meant*—because he was literally seeing and hearing and smelling what I was. By bringing a traumatically archived narrative into the present, the game simulation allowed me to reexperience it in real time instead of the artificiality of my story about it. It became real, and I have to admit, I was changed and even largely healed through the experience, which was meant only to demo the technology and not treat my own PTSD. But I never underestimated the potential of computer games again.

Computer gaming is valuable to us not just through its particular applications, but as the inkling of an approach to contending with present shock—in this case, the inability of stories to function as they used to. Without the beginnings and endings, nor the origins and goals offered by linear narratives, we must function instead in the moment. We must mourn the guiding stories we have lost, while also contending with new measures of control, freedom, and self-determination. Gaming is a great lens through which to see this process of maturation.

Narrativity is just the first of many things obsolesced by presentism, and the sense of trauma at losing linearity just the first of five main forms of present shock. But like the rest, it provokes both some initial, panicked reactions as well as a few more constructive alternatives. The disappearance of story first incites a knee-jerk sensationalism. We attempt to re-create the exhilaration and fall of traditional narrative with the increasingly lewd, provocative, or humiliating imagery of the reality-TV spectacle. Our always-on news media follows suit and, spurred further by the needs of the multinational corporations that own them, reproduce what had been the narrative authority of the newscaster with the graphic authority of the lens. How we're supposed to feel about it, on the other hand, is debated in real time as the images still play behind them.

Young people raised in this environment are among the first to take back what has been lost. Instead of finding new storytellers, they become the equivalent of storytellers themselves. Snowboarders score their own paths down a slope, while skateboarders reinterpret the urban landscape as an obstacle course. Like their peers in other pursuits, they are playing winnerless, infinite games. This growing improvisatory subculture of players also abandons the single-minded effort of political parties to win offices; they instead write their own set of behavioral norms for activism and economic justice. Instead of looking to TV and film to inform them about the world and its values, they turn to computers and games to choose their own adventures and find their own answers.

Of course, it's not as easy as all that. The digital environment presents many challenges of its own, and members of the connected generation are among the first casualties of its many distractions and discontinuities. Computers and the net may be running in real time, but its torrent of pings seems to be coming at us from all sides simultaneously. Which flashing screen we choose to answer often means less about whom or what we want to engage with than who or what we want to be, ourselves, in that moment. We're in the game, all right, but playing on many different levels at once. Or at least we're trying to.

And this struggle to be in more than one place at the same time leads to the next main type of present shock.

CHAPTER 2

DIGIPHRENIA

BREAKING UP IS HARD TO DO

Things are too busy. "An Over-Proximity of All Things." I've been fundraising. Negotiating. Flying. Every week, two plane trips, two destinations. It is all at once. No time, no time.

—SMS to me from cyborg anthropologist Amber Case

You know you're in trouble when even Google thinks you have exceeded the human capacity to exist in more than one place at a time.

It was morning, Europe-time, and I was sitting in the lobby of a hotel in a remote suburb of Berlin, jet-lagged and disoriented, unable to check in to my room until later that afternoon. Temporal limbo. But the Internet never sleeps and the lobby had WiFi, so I figured I'd get some work done. I had neglected my email for the fourteen hours it took to get from outside New York to outside

Berlin, and there were already second and third pings in the inbox from people wondering why they hadn't yet received a reply to their initial missives. (Teaches me right for answering emails so quickly and creating unrealistic expectations.)

Not that I hadn't made a valiant effort to connect while in transit. The plane had some sort of WiFi while we were over land, but someone kept trying to make Skype calls, which crashed all the passengers' connections. After the flight attendant restarted the system for the third time, I gave up, took a pill, and fell asleep to the movie. I tried again from my iPhone at the airport but couldn't log on. Now that I was at the hotel with plug-in power, I knew I'd have an easier time catching up with all that was waiting for me.

The emails with the most little red exclamation points next to them came from an executive in New York desperate to find a speaker for an event about Internet ethics. I knew I had a talk in Missouri on the day following his event but thought that maybe I could do the New York talk in the morning and still make my flight to Missouri that afternoon. I just had to log in to my Google Calendar to make sure. Google said I was using an unfamiliar IP address, so it wanted to ask a few questions to make sure I was really me. My childhood pet, my city of birth . . . that's when my thirty minutes of connectivity ran out. I went to the front desk for another little slip of paper with a password on it and got back online and back to the place where I needed to prove my identity to Google. Except now Google was really upset with me. It said I appeared to be attempting to access the site from too many locations at once. As a precaution, Google would lock my account until it could verify I was really me. Google wanted to send an automatic text message to my cell phone so I could authenticate my identity—but my cell phone did not work in Germany.

So in the haze of that rainy German morning, I foolishly decided to keep planning my digital-era schedule without access to my digital-era calendar. I replied that I would accept the morning talk as long as I could manage to leave before noon to make my flight. And as you may have guessed by now, by the time I got back to New York where I could verify my human identity to my Google media extensions, the calendar had a different version of my future than I did. My talk in Missouri was the same morning as the New York talk—not the morning after. Oops. Present-shock nightmare: I was supposed to be in two places at once. No matter how digitally adept I get, there's still only one me.

Unlike humans who might treasure their uniqueness, digital media's most defining quality is its ability to be copied exactly. Digital files, even digital processes, can be multiplied and exist in several instances at once. This is something new. The real-world, analog copies of things we have known up to now are all efforts to remain as true as possible to some original source. Even the movies we watch are struck from an internegative of the movie, which is in turn struck from the original negative. Like successive photocopies, each one contains a bit more noise—but it's better than playing the original negative repeatedly and wearing out the source. In contrast, digital copies are indistinguishable from the original. They don't take a toll on the thing being copied, either. In a sense, they are not copies at all. Once created, they *are* the original.

In the analog world, the original must be preserved, and copying slowly destroys it. Every time the original negative is run through the printing machine, it is further degraded. Flash pictures are prohibited at museums, because each exposure to light slightly dims the brilliance of an original painting. Unlike such photos, digital copies are not impressions made off a master; they are originals themselves— more like clones than children, capable of carrying on new activities as if utterly reborn in every instant.

People are still analog.

Leveraged appropriately, the immense distributive power of digital production and networks gives us the ability to spread our ideas and expressions, as well as our power and influence. We can produce effects in more than one place at a time, each of us now having the global reach formerly reserved for kings, presidents, and movie stars. Our role in our culture and society may have changed from that of passive readers or watchers to that of active game players, but this self-direction comes at a cost. We previously had the luxury of being led through experiences, our one-pointed attention moving along a tightly controlled narrative pathway. In the choose-your-own-adventure landscape of gaming—as well as the culture, workplace, and political realm structured in this new way—we must keep our eyes moving and at the same time stay aware of all sorts of activity on the periphery.

Wherever our real bodies may be, our virtual personae are being bombarded with information and missives. Our inboxes are loading, our Twitter feeds are rolling, our Facebook updates are changing, our calendars are filling, and our consumer profiles and credit reports are adjusting all along the way. As in a game, things we do not act upon don't wait for us to notice them. Everything is running in parallel, and sometimes from very far away. Timing is everything, and everyone is impatient.

Even though we may be able to be in only one place at a time, our digital selves are distributed across every device, platform, and network onto which we have cloned our virtual identities. The people and programs inhabiting each of these places relate to our digital addresses and profiles as if they were the original. The question is always "Why hasn't he answered my email?" and never "When will he log on to the Internet and check the particular directory to which my text was copied?"

Our digital devices and the outlooks they inspired allowed us to break free of the often repressive timelines of our storytellers, turning us from creatures led about by future expectations into more fully present-oriented human beings. The actual experience of this now-ness, however, is a bit more distracted, peripheral, even schizophrenic than that of being fully present. For many, the collapse of narrative led initially to a kind of post-traumatic stress disorder—a disillusionment, and the vague unease of having no direction from above, no plan or story. But like a dose of adrenaline or a double shot of espresso, our digital technologies compensate for this goalless drifting with an onslaught of simultaneous demands. We may not know where we're going anymore, but we're going to get there a whole lot faster. Yes, we may be in the midst of some great existential crisis, but we're simply too busy to notice.

We have already heard a great deal from concerned doctors and humanists about the ill effects of living digitally.[1] Their findings are on record and deserving of our consideration. While all their warnings may be true to some extent, so, too, were the warnings about automobiles and steam engines, or even the threat that written language and law posed to tribes once unified by spoken stories, and that printed Bibles posed to the authority of the Pope and his priests. The things we use do change us. In order to understand and contend with present shock, we should probably be less immediately concerned with the cause-and-effect consequences of digital activity than with the greater implications and requirements of living in the digital environment. It's not about how digital technology changes us, but how we change ourselves and one another now that we live so digitally.

We live in a world informed in large part by digital devices and outlooks, and one of the primary impacts of thinking this way is to assume the rigors of digital time as our own. Our digital universe is always-on, constantly pinging us with latest news, stock quotes, consumer trends, email responses, social gaming updates, Tweets, and more, all pushing their way to our smart phones. There are so many incoming alerts competing for attention that many phones now allow users to swipe downward to reveal a scrollable screen containing nothing but the latest alerts pushed through. Everyone and everything intrudes with the urgency of a switchboard-era telephone operator breaking into a phone call with an emergency message from a relative, or a 1960s news anchor interrupting a television program with a special report about an assassination. Anything we do may be preempted by something else. And, usually, we simply add the interruption onto the list of other things we're attempting to do at the same time.

All these interruptions, more than simply depleting our cognitive abilities, create the sense that we need to keep up with their impossible pace lest we lose touch with the present. These are live feeds, after all, pinging us almost instantaneously from every corner of the globe. There are video cameras trained on Wall Street and the Western Wall, a tent village in Cairo and a box of schnauzer puppies in a Florida pet shop.

If we could only catch up with the wave of information, we feel, we would at last be in the *now*. This is a false goal. For not only have our devices outpaced us, they don't even reflect a here and now that may constitute any legitimate sort of present tense. They are reports from the periphery, of things that happened moments ago. It seems as if to digest and comprehend them in their totality would amount to having reality on tap, as if from a fantastic media control room capable of monitoring everything, everywhere, all at the same time. It's as if all the Facebook updates, Twitter streams, email messages, and live-streamed video could combine to create a total picture of our true personal status, or that of our business, at any given moment. And there are plenty of companies out there churning all this data in real time in order to present us with metrics and graphs claiming to represent the essence of this reality for us. And even when they work, they are mere snapshots of a moment ago. Our Facebook profile and the social graph that can be derived from it, however intricate, is still just a moment locked in time, a static picture.

This quest for digital omniscience, though understandable, is self-defeating. Most of the information we get at lightning speed is so temporal as to be stale by the time it reaches us. We scramble over the buttons of the car radio in an effort to get to the right station at the right minute-after-the-hour for the traffic report. Yet the report itself warns us to avoid jams that have long since been cleared, while telling us nothing about the one in which we're currently stuck—one they'll find out about only if we ourselves call it in to their special number. The irony is that while we're busily trying to keep up with all this information, the information is trying and failing to keep up with *us*.

Meanwhile, the extraordinary measures we take to stay abreast of each minuscule change to the data stream end up magnifying the relative importance of these blips to the real scheme of things. Investors trade, politicians respond, and friends judge based on the micromovements of virtual needles. By dividing our attention between our digital extensions, we sacrifice our connection to the truer present in which we are living. The tension between the faux present of digital bombardment and the true now of a coherently living human generates the second kind of present shock, what we're calling *digiphrenia—digi* for "digital," and *phrenia* for "disordered condition of mental activity."

This doesn't mean we should ignore this digitally mediated reality altogether. For just as we found healthier responses to the fall of narrative than panic and rage, there are ways to engage with digital information that don't necessarily dissect our consciousness into discrete bits right along with it. Instead of succumbing to the schizophrenic cacophony of divided attention and temporal disconnection, we can program our machines to conform to the pace of *our* operations, be they our personal rhythms or the cycles of our organizations and business sectors. Computers don't suffer present shock, people do. For we are the only ones living in time.

TIME IS A TECHNOLOGY

We tend to think of the assault on our temporal sensibilities as a recent phenomenon, something that happened since the advent of computers and cell phones—or at least since the punch clock and shift workers. But as technology and culture theorists have reminded us at each step of the way,[2] all this started much, much earlier, and digiphrenia is just the latest stage in a very long and lamented progression. At each of these stages, what it meant to be a human being changed along with however it was—or *through* whatever it was—we related to time.

Of course, humans once lived without any concept of time at all. In this early, hunter-gatherer existence, information was exchanged physically, either orally or with gestures, in person. People lived in an eternal present, without any notion of before or after, much less history or progress. Things just were. The passage of time was not recorded or measured, but rather experienced in its various cycles. Older, wiser people and tribes became aware not just of the cycles of day and night, but of the moon and even the seasons. Since farming hadn't yet been invented, however, seasons were not to be anticipated or exploited. Beyond gathering a few nuts as it got colder, there was little we could do to shift or store time; the changes around us were simply enjoyed or endured.

Many religions and mythologies look back longingly on this prehistoric timelessness as a golden age, or Eden. Humanity is seen as a fetus in the womb, at one with Mother Nature.[3] False notions of a prehistoric noble savage aside, there is at least some truth to the idea that people lacked the capacity to distinguish themselves from nature, animals, and one another. While living so completely at the mercy of nature was fraught with pain and peril, this existence was also characterized by a holism many media and cultural theorists consider to be lost to us today in a world of dualism, preferences, and hierarchies. As media theorist and Catholic priest Walter Ong put it, "Oral communication unites people in groups. Writing and reading are solitary activities that throw the psyche back on itself. . . . For oral cultures, the cosmos is an ongoing event with man at its center."[4] People living in this oral, timeless civilization saw God, or the gods, in everything around them. While they had to worry about where their next meal was coming from, they felt no pressure to succeed or to progress, to achieve or to improve. They had nowhere to go, since the very notion of a future hadn't yet been invented. This stasis lasted several thousand years.

Everything changed, finally, in the Axial Age with the invention of text. The word-pictures of hieroglyphic writing were replaced with the more discrete symbols of an alphabet. The progenitor of a more digital style of storage, letters were precise and abstract. Combined together, they gave people a way to represent the mouth noises of oral culture in a lasting artifact. Like a digital file, a spelled word is the same everywhere it goes and does not decay. The simple twenty-two-letter alphabet popularized and democratized writing, giving people a way to record promises, debts, thoughts, and events. The things written down one day could be read and retrieved the next.

Once a line could truly be drawn in something other than sand, the notion of history as a progression became possible. With the invention of text came the ability to draft contracts, which were some of the first documents ever written, and described agreements that endured over time. With contracts came accountability, and some ability to control what lay ahead. The notion of a future was born. Religion, in the oral tradition, came from the mouth of a leader or pharaoh, himself a stand-in for God. Text transformed this passive relationship to God or nature with a contract, or, more precisely, a covenant between people and God. What God demands was no longer a matter of a tyrant's whim or the randomness of nature, but a set of written commandments. Do this and you will get that.

This resonated well with people who were learning agriculture and developing a "reap what you sow" approach to their world. Seeds planted and tended now yield a crop in the future. Scriptural laws obeyed now earn God's good graces in the future. The world was no longer just an endless churn of cycles, but a place with a past and a future. Time didn't merely come around; it flowed more like a river, forming a history of all that went before. In the new historical sense of time, one year came after the other. Human beings had a story that could be told—and it was, in the Torah and other written creation myths. Pagan holidays that once celebrated only the cycle of the seasons now celebrated moments in history. The spring equinox and fertility rites became the celebration of the Israelite exodus from Egypt; the solstice became the Hanukkah reclamation of the Temple and, later, the birth of Jesus. Periods in the cycle of nature became moments in the flow of history.[5]

The new metaphor for time was the calendar. A people was defined and its activities organized by its calendar, its holidays, and its memorials. Calendars tell a culture what matters both secularly and religiously. The time for sacred days was held apart, while time for productivity could be scheduled and even enforced. The calendar carried the double-duty of representing the cyclical nature of the lunar months and solar year while also keeping track of historical time with the passing of each numbered year. There was now a before and an after—a civilization that could measure its progress, compare its bounties from one year to the next, and, most important, try to do *better*. The great leaning forward had begun. We progressed from what social theorist Jeremy Rifkin called "the Earth's universe" to "God's universe,"[6] conceiving ourselves as participants in a greater plan and subject to a higher law and an external gauge of our success over time.

Some of the most devout members of this religious universe were responsible for breaking time down even further. For their new Islamic faith, Muslims were required to pray at regular intervals. Their methodical call to prayer sequence used the height of the sun and measurement of shadows to break the day into six sections. In Europe, it was Benedictine monks who organized not just the calendar year but every day into precisely defined segments for prayer, work, meals, and hygiene. Handheld bells coordinated all this activity, making sure the monks performed their tasks and said their prayers at the same time. Surrendering to this early form of schedule constituted a spiritual surrender for the medieval monks, for whom personal time and autonomy were anathema to their new collective identity. Although their schedule might look simple compared with that of an average junior high student today, the monks were exercising radically strict temporal discipline for the time.

As they became more concerned (some may argue obsessed) with synchronizing all their daily routines, the monks eventually developed the first mechanical timepieces. The Benedictine clocks were celebrated for their escapement technology—basically, the ability to control the descent of a weight (or expansion of a spring) by breaking its fall slowly and sequentially with a little ticking gear. The real leap, however, had less to do with escapement than with ticking and tocking itself. What the monks had discovered was that the way to measure time was to break it down into little beats. Just as ancient Buddhist water clocks could mark four hours by storing the combined volume of hundreds of relatively regularly falling drops, the Benedictine clocks broke down the slow, continuous descent of weights into the regular beats of a pendulum. Ticktock, before-after, yes-no, 1/0. Time was necessarily digital in character,[7] always oscillating, always dividing. As an extension of the new culture of science (a word that originally meant to separate one thing from another, to split, divide, dissect), the clock turned time into something that divides, and, like any technology, created more preferences, judgments, and choices.

Even though the Chinese had accurate water clocks for centuries before the Benedictines, clocks and timing did not come to spread and dominate Asian culture the same way they did in Europe. Westerners believed this was because the Chinese didn't know quite what to do with all this precision. But it may have had less to do with a lack than with a bounty. The Chinese already had a strong sense of culture and purpose, as well as a different relationship to work and progress over time. The introduction of timepieces capable of breaking down time didn't have quite the same impact on a people who looked at time—for better and for worse—as belonging to someone else, anyway.

Arriving on church bell towers at the dawn of the Industrial Age, the clock was decidedly more interesting to those looking for ways to increase the efficiency of the new working classes. Ironically, perhaps, an invention designed to affirm the primacy and ubiquity of the sacred ended up becoming a tool for the expansion of the secular economy. Trade had been expanding for a century or two already, and keeping track of things numerically—as well as temporally—had become much more important. If the previous era was characterized by the calendar, this new clockwork universe would be characterized by the schedule.

The bells of the monastery became the bells of the new urban society. Trade, work, meals, and the market were all punctuated by the ringing of bells. In line with other highly centralizing Renaissance inventions such as currency and the corporation, bells were controlled by central authorities. This gave rise to distrust, as workers were never sure if their employers were measuring time fairly. The emergence of the clock tower gave everyone access to the same time, allowing for verification while also amplifying time's authority.

Thanks to the clock tower, the rhythms of daily life were now dictated by a machine. Over time, people conformed to ever more precisely scheduled routines. Where the priority of the calendar-driven civilization was God, the priorities of the clockwork universe would be speed and efficiency. Where calendars led people to think in terms of history, clocks led people to think in terms of productivity. Time was money. Only after the proliferation of the clock did the word "speed" (spelled *spede*) enter the English vocabulary, or did "punctual"—which used to refer to a stickler for details—come to mean a person who arrived on time.[8]

The metaphor for the human being became the clock, with the heartbeat emulating the ticks of the escapement, counting off the seconds passing. Management of people meant management of time (the word "management" itself deriving from putting a horse through its paces, or *manege*). People were to perform with the precision and regularity of the machines they drove—and, in some senses, were becoming. By the 1800s, workers punched clocks to register their hours. A mechanical engineer named Frederick Taylor applied his skill with machines to human beings, inventing a new field called scientific management. He and his assistants would spread out through a company armed with stopwatches and clipboards to measure and maximize the efficiency of every aspect of the work cycle. The time it took to open a file drawer was recorded down to the hundredth of a second, in order to determine the standard time required to complete any job. Once that was known, the efficiency of any particular worker could be measured against it. The efficiency movement was born, for which glowing accounts of increased productivity over time were published and promoted, while evidence of worker dissent was actively suppressed.[9]

Now that human beings were being tuned up like machines, the needs of humans and machines became almost indistinguishable. The entirety of the clockwork universe may as well have been a machine, with new innovations emerging primarily to assist technology or the businesses on which those technologies depended. Thanks in part to the legal arguments of a railroad industry lawyer named Abraham Lincoln, for example, the rights of local municipalities were subordinated to those of corporations that needed thoroughfare for their trains and cargo. Time and timing began to mean more than place. Transcontinental commerce required synchronized activity over great distances, leading to "standard time" and the drawing of time zones across the map. (Greenwich Mean Time's placement in the United Kingdom represented the British Empire's lingering domination of the globe.) Likewise, the telegraph emerged primarily as a communication system through which train crashes could be minimized. Directing the motion of trains with red lights and green lights was eventually applied to cars and ultimately to people navigating the crosswalks—all timed to maximize efficiency, productivity, and speed. In the clockwork universe, all human activity—from shift work to lunch breaks to TV viewing to blind dates—involved getting bodies to the right place at the right time, in accordance with the motions of the clock. We were as clocks ourselves, with arms that moved and hearts that counted and alarms that warned us and bells that went off in our heads. Just wind me up in the morning.

If the clockwork universe equated the human body with the mechanics of the clock, the digital universe now equates human consciousness with the processing of the computer. We joke that things don't compute, that we need a reboot, or that our memory has been wiped. In nature, our activities were regulated by the turning of the Earth. While the central clock tower may have coordinated human activity from above, in a digital network this control is distributed—or at least it seems that way. We each have our own computer or device onto which we install our choice of software (if we're lucky), and then use or respond to it individually. The extent to which our devices are conforming to external direction and synchronization for the most part remains a mystery to us, and the effect feels less like top-down coordination than personalized, decentralized programs.

The analog clock imitated the circularity of the day, but digital timekeeping has no arms, no circles, no moving parts. It is a number, stationary in time. It just is. The tribal community lived in the totality of circular time; the farmers of God's universe understood before and after; workers of the clockwork universe lived by the tick; and we creatures of the digital era must relate to the pulse. Digital time does not flow; it flicks. Like any binary, discrete decision, it is either here or there. In contrast to our experience of the passing of time, digital time is always in the now, or in no time. It is still. Poised.

I remember when I was just ten years old, how I used to stare at my first digital clock. It had no LED, but rather worked a bit like a train terminal's board, with a new number flipping down into place every minute. I would wait and count, trying and failing to anticipate the click of the next number flipping down—each time being surprised by its suddenness in a micromoment of present shock. My dad's old alarm clock required him to wind it up each night, and then to twirl a second winder for the alarm. Over the course of the day, the potential energy he wound into the device slowly expressed itself in the kinetic energy of the motion of the arms and bell hammer. My digital clock just sat there, interrupting itself each minute only to sit there again. Its entire account of the minute 7:43 was the same. Maybe that's why, to this day, digital watches have not replaced their analog counterparts—though wristwatches are primarily ornamental for most wearers, who now read the time on a smart phone.

That is, if the smart phone only sat there and waited for us to read it. More often than not, it's the phone or laptop demanding *our* attention, alerting us to the upcoming event in our schedule, or unpacking one of a seemingly infinite number of its processes into our attention. Indeed, if the Axial Age was coordinated by the calendar, and the clockwork universe by the schedule, the digital era subjects us to the authority of code. Our children may have their afternoons scheduled, but we adults live in a world that is increasingly understood as a program.

Where a schedule simply lists an appointment we are supposed to attend or a task we are to complete, a program carries out the actual process with us or for us. The clock dictated only time; the computer (and smart phone, and biometric device on the wrist) tells us what we should be doing when. We once read books at our own pace; computer animations, YouTube movies, and video games play out at a speed predetermined by the programmer. The GPS on the dashboard knows our entire route to the dentist's office and offers us the turns we need to take on a need-to-know basis. The computer graphic on the exercise machine tells us how fast to pedal and turns the simulation uphill when it knows we need it. We surrender this authority over our pacing because we know that, for the most part, the programs will be better at helping us meet our stated goals than we are. The events carried out by computers are less like schedules than they are like simulations; less like sheet music, more like a piano roll. In a scheduled world, you are told you have half an hour to peruse an exhibit at the museum; in a programmed world, you are strapped into the ride at Disneyland and conveyed through the experience at a predetermined pace. However richly conceived, the ride's complexity is limited because its features are built-in, appearing and unfolding in sync with the motions of your cart. The skeleton dangles, the pirate waves, and the holographic ship emerges from the cloud of smoke. In the programmed life, the lights go on and off at a specified time, the coffee pot activates, the daylight-balanced bulb in the programmable clock radio fades up. Active participation is optional, for these events will go on without us and at their own pace.

As we grow increasingly dependent on code for what to do and when to do it, we become all the more likely to accept a model of human life itself as predetermined by the codons in our genomes.

All we can do is watch the program of our DNA unfold or, at best, change our fates by altering the sequence of genes dictating what's supposed to happen. Free will and autonomy are eventually revealed to us to be a simulation of some sort, while the reality we think we're participating in is really just the predetermined dance of pure information. The timekeeper is no longer the controller of the clock but the programmer of the computer.

And instead of taking our cues from the central clock tower or the manager with the stopwatch, we carry our personal digital devices with us. Our daily schedule, dividing work time from time off, is discarded. Rather, we are *always*-on. Our boss isn't the guy in the corner office, but a PDA in our pocket. Our taskmaster is depersonalized and internalized—and even more rigorous than the union busters of yesterday. There is no safe time. If we are truly to take time away from the program, we feel we must disconnect altogether and live "off the grid," as if we were members of a different, predigital era.

Time in the digital era is no longer linear but disembodied and associative. The past is not something behind us on the timeline but dispersed through the sea of information. Like a digital unconscious, the raw data sits forgotten unless accessed by a program in the future. Everything is recorded, yet almost none of it feels truly accessible. A change in file format renders decades of stored files unusable, while a silly, forgotten Facebook comment we wrote when drunk can resurface at a job interview.

In the digital universe, our personal history and its sense of narrative is succeeded by our social networking profile—a snapshot of the current moment. The information itself—our social graph of friends and likes—is a product being sold to market researchers in order to better predict and guide our futures. Using past data to steer the future, however, ends up negating the present. The futile quest for omniscience we looked at earlier in this chapter encourages us, particularly businesses, to seek ever more fresh and up-to-the-minute samples, as if this will render the present coherent to us. But we are really just chasing after what has already happened and ignoring whatever is going on now. Similarly, as individuals, our efforts to keep up with the latest Tweet or update do not connect us to the present moment but ensure that we are remaining focused on what just happened somewhere else. We guide ourselves and our businesses as if steering a car by watching a slide show in the rearview mirror. This is the disjointed, misapplied effort of digiphrenia.

Yet instead of literally coming to our *senses*, we change our value system to support the premises under which we are operating, abstracting our experience one step further from terra firma. The physical production of the factory worker gives way to the mental production of the computer user. Instead of measuring progress in acres of territory or the height of skyscrapers, we do it in terabytes of data, whose value is dependent on increasingly smaller units of time-stamped freshness.

Time itself becomes just another form of information—another commodity—to be processed. Instead of measuring change from one state of affairs to another, we measure the rate of change, and the rate at which that rate is changing, and so on. Instead of proceeding from the past to the future, time now proceeds along derivatives, from location to speed to acceleration and beyond. We may like to think that the only constant is change, except for the fact that it isn't really true—change is changing, too. As Mark McDonald, of IT research and advisory company Gartner, put it, "The nature of change is changing because the flow and control of information has become turbulent, no longer flowing top down, but flowing in every direction at all times. This means that the ability to manage and lead change is no longer based on messaging, communication and traditional sponsorship. Rather it is based on processes of informing, enrolling and adapting that are significantly more disruptive and difficult to manage for executives and leaders."[10]

Or as Dave Gray, of the social media consultancy Dachis Group, explains it, "Change is not a once-in-a-while thing so much as something that is going to be happening all the time. Change is accelerating, to the point where it will soon be nearly continuous. Periods of sustained competitive advantage are getting shorter, and there are a host of studies that confirm that. It's not just something that is happening in technology, either. It's happening in every industry."[11]

These analysts are describing the new turbulence of a present-shock universe where change is no longer an event that happens, but a steady state of existence. Instead of managing change, we simply hope to be iterated into the next version of reality that the system generates. The only enduring truth in such a scheme is evolution, which is why the leading spokespeople for this world-after-calendars-and-clocks tend to be evolutionary scientists: we are not moving through linear time; we are enacting the discrete, punctuated steps of a program. What used to pass for the mysteriousness of consciousness is shrugged off as an emergent phenomenon rising from the complexity of information. As far as we know, they may be right.

It's not all bad, of course. There are ways to inhabit and direct the programs in our lives instead of simply responding to their commands. There are ways to be in charge. Unlike the workers of the Industrial Age who stood little chance of becoming one of the managing elite, we are not excluded from computing power except through lack of education or the will to learn. As we'll see, becoming one of the programmers instead of the programmed is probably a better position from which to contend with digitality.

CHRONOBIOLOGY

Thanks to our digital tools, we are living in a new temporal order, one no longer defined by the movement of the heavens, the division and succession of the years, or the acceleration of progress. We are free of the confines of nature, capable of creating simulated worlds in which we can defy gravity, program life, or even resurrect ourselves. And as anyone who has gotten lost in World of Warcraft only to look up at the clock and realize four hours have gone by can tell you, we are also free of time—and as cognitive studies show, more so than when reading a book or watching a movie, and with temporal distortions lingering still hours later.[12]

At least our virtual selves enjoy this freedom. We flesh-and-blood humans living back in the real world have still aged four hours, missed lunch, denied ourselves bathroom breaks, and allowed our eyes to dry up and turn red. Like an astronaut traveling at light speed for just a few seconds who returns to an Earth on which ninety years have passed, our digital selves exist in a time unhinged from that of our bodies. Eventually the two realities conflict, leading to present shock. If tribal humans lived in the "total" time of the rotating Earth, digital humans attempt to live in the "no" time of the computer. We simply can't succeed at it if we bring our bodies along for the ride. Yet when we try to leave them behind, both nature and time come back to reassert their authority over us.

Digital technology is not solely to blame. Indeed, the microchip may be less the cause of this effort to defeat time than its result. Since the prehistoric age, humankind has been using technology to overcome the dictates of nature's rhythms. Fire allowed us to travel to colder climates and defeat the tyranny of the seasons. It also gave us the ability to sit up past sundown and cook food or tell stories. In the early 1800s, the proliferation of gaslight utterly changed the landscape and culture of London, making the night streets safer and illuminating the urban environment at all hours. New cultures emerged, with new relations to time of day and other activities. Nighttime cafés and bars led to new musics and entertainments. New flexibility of work scheduling allowed for around-the-clock shifts and factories whose stacks emitted smoke day and night. The invention of jet planes gave us even more authority over time, allowing us to traverse multiple time zones in a single day.

But no matter how well technology overcame the limits of natural time, our bodies had difficulty keeping up. Stress and fatigue among night workers and those whose shifts are changed frequently is only now being recognized as more than a ploy by labor unions, and if it weren't for the rather obvious symptoms of jet lag, we may still not have acknowledged the existence of biological clocks underlying our bodies' rhythms. For jet lag is more than just a woozy feeling, and it took many years until scientists realized it had a real influence over our effectiveness. Back in the 1950s, for example, when jet passenger service was still quite novel, Secretary of State John Foster Dulles flew to Egypt to negotiate the Aswan Dam treaty. His minders assumed he would sleep on the plane, and they scheduled his first meeting for shortly after he arrived. He was incapable of thinking straight, his compromised perceptual and negotiating skills were overtaxed, and he failed utterly. The USSR won the contract instead, and many still blame this one episode of jet lag for provoking the Cold War.

A decade later, in 1965, the FAA finally began to study the effects of air travel on what were now admitted to be human biological rhythms. For some unknown reason, subjects traveling west to east experienced much greater decline in "psychological performance" than those traveling in other directions.[13] The next year, the *New York Times* Sports section acknowledged the little-understood effects of jet lag on Major League Baseball players: "The Jet Lag can make a recently debarked team too logy for at least the first game of any given series."[14] Coaches became aware of the various behavior patterns associated with travel in each direction, but no one could tell them the mechanisms at play or how to counteract them.

By the 1980s, NASA got on the case. Its Ames Fatigue/Jet Lag Program was established to "collect systematic, scientific information on fatigue, sleep, circadian rhythms, and performance in flight operations."[15] Leaving people in rooms with no external time cues, researchers found that the average person's biological clock would actually lengthen to a twenty-five-hour cycle. This, they concluded, is why traveling east, which shortens the day, is so much more disorienting than traveling west, which lengthens it. Most important, however, the studies showed that there were clocks inside us, somehow governed by the body, its metabolism, and its chemical processes. Or perhaps we were syncing to something unseen, like the moon, or shifting magnetic fields. Or both. Circadian rhythms, as they came to be called, were real.

The phenomenon had been discovered and measured in plants centuries earlier. In the 1700s, Swedish botanist Carolus Linnaeus designed a garden that told time by planting adjacent sections of flowers that opened and closed their blossoms an hour apart, around the clock. But if human activities were really governed by the same mysterious cycles as tulips and cicada, we would have to consider whether our ability to transcend nature through technology was limited by forces beyond our control.

The relatively new field of chronobiology hopes to unravel some of these mysteries, but each new discovery seems to lead to even bigger questions. Some biological cues are clearly governed by simple changes in sunlight. The eye's photoreceptors sense the darkening sky, sending a signal to release melatonin, which makes us sleepy. Watching TV or staring at a bright computer screen in the evening delays or prevents this reaction, leading to sleeplessness. But if sunlight were the only cue through which the body regulated, then why and how does a person who can set his own hours find the twenty-five-hour day?

It is because we also have internal clocks, governed by less understood metabolic, hormonal, and glandular processes. We listen to those inner rhythms while simultaneously responding to external cues, from daylight and moon phases to the cycle of the seasons. Still other evidence suggests a complex set of relationships between all these clocks. Body temperature rises during the daylight hours, but some people's rise faster than others, making them morning people, while those whose temperature rises more slowly over the course of the day reach their most effective state of consciousness in the early evening. Meanwhile, when our body temperature rises, we perceive of time as passing more slowly. This is because our internal clocks are counting faster, while time is actually passing at the same rate as always.

Although the thing we call time might be a mere concept— some variation on energy in Einstein's equations—all this chrono-biological evidence suggests there is a kind of synchronization going on between different parts of our world. In other words, even if we are ultimately unhinged from any absolute clock at the center of the universe, we are not unhinged from one another. In a biosystem or a culture, timing is everything.

For instance, the daily orbit of the moon generates two high tides every twenty-four hours. Twice each month, at the full and new moons, the tide is higher than normal, creating the fifteen-day-cycle spring tides that dictate the life and mating cycles of marine life. The human menstrual cycle is approximately twenty-eight days, the same cycle as the moon, for reasons and through mechanisms no one has yet figured out. In other chronobiology research, we learn that people treated with chemotherapy at one hour of the day respond significantly better than those treated at another hour.

The point is that time is not neutral. Hours and minutes are not generic, but specific. We are better at doing some things in the morning and others in the evening. More incredible, those times of day change based on where we are in the twenty-eight-day moon cycle. In one week, we are more productive in the early morning, while in the next week we are more effective in the early afternoon.[16]

Technology gives us the ability to ride roughshod over all these nooks and crannies of time. We can fly through ten time zones in as many hours. We can take melatonin or Ambien to fall asleep when we've arrived at our destination, and later take one of our attention deficit disorder–afflicted son's Ritalin pills to wake up the next morning. Then maybe get a prescription for Prozac or Lexapro to counteract the depression and anxiety associated with this lifestyle, and one for a good tranquilizer to calm the mind at night with all those mood enhancers in our bloodstream, and maybe one for Viagra to counteract the sexual side effects.

We can screw up our biological clocks a lot easier than that, too. Shift work, where employees alternate between days and nights on the job, leads to a significantly higher rate of violence, mood disorders, depression, and suicide. If a shift worker is scheduled even just one night on duty, urinary electrolytes take five days to adjust and eight days for the heart rate to return to normal. The World Health Organization has suggested that shift work is a "possible" carcinogen.[17] Women who work night shifts, for example, may have up to a 60 percent greater risk of contracting breast cancer. Even that midnight snack cues your body to believe it's daytime, reducing the effectiveness of your sleep.

Technology gives us more choice over how and when we do things. What we often forget is that our bodies are not quite as programmable as our schedules. Where our technologies may be evolving as fast as we can imagine new ones, our bodies evolved over millennia, and in concert with forces and phenomena we barely understand. It's not simply that we need to give the body rhythms; we can't simply declare noon to be midnight and expect the body to conform to the new scheme. If only it were as simple as a single clock for which we could change the settings—but it's not. The body is based on hundreds, perhaps thousands, of different clocks, all listening to and relating to and syncing with everyone and everything else's. Human beings just can't evolve that quickly. Our bodies are changing on a much different timescale.

PACING AND LEADING

Luckily, our technologies and programs are as fungible as our bodies are resistant. Yes, we are in a chronobiological crisis of depression, suicides, cancers, poor productivity, and social malaise as a result of abusing and defeating the rhythms keeping us alive and in sync with nature and one another. But what we are learning gives us the ability to turn this crisis into an opportunity. Instead of attempting to retrain the body to match the artificial rhythms of our digital technologies and their artifacts, we can instead use our digital technologies to reschedule our lives in a manner consistent with our physiology.

Technology may have given us the choice to defeat our natural rhythms, but we then built a society and economic order around these choices, making them seemingly irreversible. The only answer seemed to be to speed up, and the microchip might well be the poster child for this race to catch up with ourselves. If only we could offload the time-intensive tasks to silicon, we would regain the time—what technology analyst Clay Shirky calls "cognitive surplus"[18]—that we need to do our thinking again.

But our microchips don't seem to be serving us in that way. Instead of our offloading time-intensive tasks to our machines, our machines keep us humans working at their pace, or the pace of the companies on the other end of our network connections. Thanks to the Internet, we travel more on business, not less, we work at all hours on demand, and we spend our free time answering email or tending to our social networks. Staring into screens, we are less attuned to the light of day and the physiological rhythms of our housemates and coworkers. We are more likely to accept the false digital premise that all time is equivalent and interchangeable.

The always-on philosophy works well for many businesses. During overnight hours when traditional broadcast television networks used to go off the air, cable channels like Home Shopping Network and QVC are still buzzing. Where traditional department stores may wait until the fourth quarter of the year for one of the store's buyers to walk the floor and find out how certain product lines are doing, executives at QVC get sales reports as they happen. "It's instant gratification," explained one merchandiser using QVC to push product. "You learn and react constantly."[19]

Their salespeople sure do. Mostly former (or, one could argue, still-working) actors, these shopping-channel salespeople make pitches for zirconium diamonds and designer handbags while receiving constant, real-time feedback on their success. Computer screens indicate the number of items sold, how many remain in inventory, and whether the pace of sales is increasing or decreasing—allowing the announcer to shift strategies and intensity instantaneously. They also receive feedback through an earpiece, about everything from the quality of their pitch, merchandiser satisfaction with the on-air delivery style, and whether phone lines are jammed. If a sweater is selling best in red, the host sees this on the screen, then picks up the green one and talks about its beautiful shade, or how "only fifty are left in stock. Make that forty-five—better hurry!" The salespeople almost universally feel that these technological links allow them to forge an emotional connection with their audiences, even though most of the feedback they get is in numbers. Rick Domeier, a ten-year veteran salesman, explains, "If you go too hard, they'll let you know." Their skill at reading and digesting multiple sources of information on the fly has led many to compare them with air traffic controllers. As far as the hosts are concerned, they are living outside time: "This is kind of like Vegas—you don't know if it's two o'clock in the afternoon or two o'clock in the morning," adds Domeier. "When I'm on, it's prime time."[20]

Many of us aspire to this ability to be "on" at any time and to treat the various portions of the day as mere artifacts of a more primitive culture—the way we look at seemingly archaic blue laws that used to require stores to remain closed at least one day a week. We want all access, all the time, to everything—and to be capable of matching this intensity and availability ourselves. Isn't this what it would mean to be truly digital citizens of the virtual city that never sleeps? A cell phone can be charged at any hour of the day, just as a person can take a power nap. Ultraefficiency advocate Timothy Ferriss's book *The 4-Hour Body* teaches readers how to "hack sleep." Rather than sleeping six to nine hours in one stretch, those practicing polyphasic sleep are supposed to get by with twenty-minute naps every four hours. It's an approach to the human-body-as-lithium-ion-battery that Ferriss says has proved successful for several high-tech CEOs.[21] It's also indicative of a set of aspirations that may be more appropriate for machines, or markets, than for people.

Instead of demanding that our technologies conform to ourselves and our own innate rhythms, we strive to become more compatible with our technologies and the new cultural norms their timelessness implies. We compete to process more emails or attract more social networking connections than our colleagues, as if more to do on the computer meant something good. We misapply the clockwork era's goals of efficiency and productivity over time to a digital culture's asynchronous landscape. Instead of working inside the machine, as we did before, we must *become* the machine.

We fetishize concepts such as the cyborg or human technological enhancement, looking to bring our personal evolution up to the pace of Apple system updates. We answer our email in the morning before we brush our teeth,[22] we trust the computers at Echemistry.com to calculate our best mates, and we line up for new iPhones even though our old ones still work. We obsessed over the first day a social networking website called Facebook sold its stock on the NASDAQ exchange, which had less to do with any interest in the value of the stock than an acknowledgment of social networking as central to our lives: the online network as the default mode of human connectivity and one's profile as the new form of self-representation as well as the mirror in which to primp. Our algorithmically generated Klout scores stand in for what used to be social status. The NikeFuel bands strapped around our wrists track our motions and change colors as we approach our fitness goal for the day, then upload our results with other cyber-enhanced exercisers in our network, for mutual motivation.

Some of this activity happens so self-consciously that it amounts to what digital-arts theorists call the New Aesthetic[23] of everything from drone photography and ubiquitous surveillance to 8-bit game nostalgia and pixilated herringbone fabric. It is a blending of the language of digital technology with that of the physical world, perhaps best exemplified by the most advanced form of dubstep dancing, called glitch. Taking up where robot dancers and pop-and-lockers left off, glitch dancers imitate the glitchy stutter of low-resolution video streaming over the Internet. Their movements imitate those of a dancer as rendered by a malfunctioning video device, complete with dropouts and stutters.

Internet workers are expected to accept the cyborg ethos as a given circumstance. Google and Facebook welcome their engineers to work around the clock; the companies provide food, showers, and even laundry service for their programmers—who are assumed to have no life beyond the company. The bathroom stall doors at Google have daily programming tips for employees to read while sitting on the toilet. These campuses are lovely, to be sure, and the food and facilities are of a higher standard than most of us get at home. And everyone appears to be having a good time. But these places may as well be space stations, meticulously stocked and arranged, and utterly cut off from the passage of time.

The culture of the Internet is informed by this sensibility, which then trickles back into its content and programs. Programmers who are always-on naturally assume their users would want to live this way as well. Likewise, Internet writers whose lives must conform to the dictates of online publishing and economics end up espousing values consistent with the always-on hyperactivity of the Web. The most trafficked sites have new posts going up every few minutes, and their writers feel the pressure. Blog publisher Nick Denton has grown famous—and wealthy—exploiting this phenomenon in the way he runs his own properties, Gawker, Gizmodo, and Jezebel, among others. The formula is to hire young writers, subject them to almost impossible time constraints, and then let them go if they develop a reputation and merit higher pay. Writers for Denton's publications are required to post twelve pieces per day—which often keeps them working around the clock (or, as I have witnessed myself, posting via smart phones from "real" life activities). The blog's content exudes this sense of desperate immediacy and always-on urgency. No celebrities are safe or should be safe, wherever and whenever they are. A reputation-busting story about almost anyone can break while he or she is asleep, and spread all over the net before morning. As if to respond to the culture these blogs created, Denton implemented a computer program at Gawker that assigned stories based on trending Internet terms, creating a closed feedback loop between the net's writers and readers.

It's a lifestyle that is at once endured and enjoyed. As *New York* magazine writer and Gawker confidante Vanessa Grigoriadis put it,

> Bloggers get to experience the fantastic feeling of looking at everything in the world and then having everyone look at them through their blog, of being both subject and object, voyeur and voyeurant. To get more of that feeling, some bloggers—if we were a blog, we'd tell you who—are in the bathroom snorting cocaine, or Adderall, the ADHD drug popular among college kids on finals week, the constant use of which is one of the only ways a blogger can write that much ("We're a drug ring, not a

bunch of bloggers," one Gawker Media employee tells me cheerily). Pinched nerves, carpal tunnel, swollen feet—it's all part of the dastardly job, which at the top level can involve editing one post every fifteen minutes for nine hours a day, scanning 500 Websites via RSS for news every half-hour, and on "off-hours" keeping up with the new to prepare for tomorrow.[24]

By letting technology lead the pace, we do not increase genuine choice at all. Rather, we disconnect ourselves from whatever it is we may actually be doing. Bloggers disconnect themselves from the beats they may be covering by working through the screen and keyboard, covering the online versions of their subjects. Designers base their fashions and handbags on the computer readouts of incoming calls from housewives at 1 a.m. Lovers expect immediate and appropriate responses to their text messages, however tired or overworked (or drunk) the partner might be. Programmers expect themselves to generate the same quality code at 2 a.m. as they did at 2 p.m. earlier— and are willing to medicate themselves in order to do so.

In each of these cases, the bloggers, designers, lovers, and programmers all sacrifice their connection to natural and emergent rhythms and patterns in order to match those dictated by their technologies and the artificial situations they create. They miss out on the actual news cycle and its ebb and flow of activity. They work less efficiently by refusing to distinguish between naturally peak productive and peak restorative hours. Designers miss out on quite powerfully determinative cultural trends and cycles by focusing on the mediated responses of insomniac television viewers. And their articles, programs, and creative output all suffer for it.

It's an easy mistake to make. The opportunity offered to us by digital technology is to reclaim our time and to program our devices to conform to our personal and collective rhythms. Computers do not really care about time. They are machines operating on internal clocks that are not chronological, but events-based: *This* happens, then *that* happens. They don't care how much—or how little—time passes between each step of the sequence. This relationship to time offers unique opportunities.

For example, early Internet culture took advantage of the staccato, stepped no-time of digital technology. The first bulletin board services worked as asynchronously as the computer programs themselves: users logged on, downloaded entire conversations, and then responded at their convenience. A participant might take a few hours, or even a few days, to craft a response to a conversation in progress, and then upload it to the bulletin board. We conversed and wrote at our own pace and in our own time. For some this meant reading at night and then composing a response in the morning. For others, it meant a sudden burst of reading and writing. For still others like me, it took on a quality of chess by mail, with responses going back and forth over a period of days.

The beauty of these conversations is that we were all brilliant—more brilliant than we were in real life, anyway, when only the very best conversationalists were capable of coming up with replies that took the rest of us days to construct. The result was that on the Internet we were smarter than we could be in daily life. We had used digital technology not simply to slow down a group's conversation, but to allow everyone to participate at his or her own pace and in the most effective part of his or her daily or weekly cycle.

By putting email and Twitter in our smart phones and attaching them to our bodies so that something vibrates every time we are mentioned, summoned, or pinged, we turn a potentially empowering asynchronous technology into a falsely synchronous one. We acquire phantom vibration syndrome and begin to experience symptoms formerly limited to air traffic controllers and 911 emergency operators. Even when we take advantage of speed or, pardon, "cognitive enhancement drugs," we eventually burn out. The same is true for businesses that rework their operations around the time frames of quarterly reports, and mutual funds that "dress up" for end-of-year evaluations by selling unpopular assets and purchasing popular ones. They miss or work against more naturally occurring rhythms and lose the ability to ride the waves underlying essentially all human endeavors. The phones are smarter but we are dumber.

Countless examples of self-pacing systems abound in both nature and culture. The mentrual cycles of women who live in the same dormitory tend to become synchronized. Through a perhaps similar mechanism, children's playground activity has been observed to be paced by just one or two youngsters who move about seemingly randomly from one group to another, establishing fairly precise rhythms of activity. Social scientists believe this collaborative pacing allows for high levels of group coherence as well as greater awareness of those individuals falling out of sync due to illness or some other stress deserving of collective attention. These processes may be little more than artifacts of more primitive human societies, but they may also be essential coordinators of activity and behavior that we should think twice about overriding.

Rather than being paced by our technologies, we can just as easily program our technologies to follow our own paces—or those of our enterprise's remaining natural cycles. Or better than simply following along, technologies can sync to us and generate greater coherence for all of us in the process. After all, people have been achieving the benefit of sync since the invention of agriculture. Farmers learned that certain crops grow better in particular climates and seasons, so they plant the right seeds at the right times. Not only is the crop better and more bountiful when planting is organized in this fashion, but the fruits, vegetables, and grains available end up better matched to the human physiology's needs during that season. Potatoes, yams, carrots, beets, and other root vegetables are in high supply during the winter months, providing sustained energy and generating warmth. The moist, hydrating fruits available during the summer are cooling to the overheated body. Beyond these superficial relationships are very specific glandular and hormonal connections between seasonal shifts and available plant enzymes. Whatever is in season has been in season over the course of hundreds of thousands of years of human evolution, coaxing and cueing everything from our thyroids to our spleens to store, cleanse, and metabolize at appropriate intervals. We can willfully re-create the dietary regimen that the year-round availability of hydroponic vegetables in the aisles of the Whole Foods Market otherwise effectively camouflages from us.

Much of our culture is already designed around the more subtle influences of the seasons on our moods, hormone levels, and neurotransmitter activity. Religions program their holidays to exploit or, in some cases, counteract seasonal states of mind. Ancient fertility rituals came in April, when both farming and sexual activity reached their heights, redirecting it to local gods and authorities. Solstice rituals brought light, greenery, or oily foods into people's homes during the darkest days of the year, when as-yet-unidentified seasonal affective disorder was most likely to strike. Today it's no secret that movie studios release blockbuster action films to meet the higher energy levels of summer audiences, more intellectual fare for the winter months, and romantic comedies for spring. They don't do it out of any sense of loyalty to our natural chronobiological rhythms, but because it's good business.

People can program their own activities to conform to these cycles as well—once we know what they are. Athletes and their trainers have always been aware that surges of energy, high performance levels, and unexplained slumps seemed to follow regular cycles over seasons, months, weeks, and even times of day. But it wasn't until the 1970s that a surging culture of amateur joggers looking mostly for better ways to improve their cardiac health made the first significant breakthroughs tying timing to performance.

Already accustomed to making minute tweaks to their highly routinized workout regimens, runners took a particularly keen interest in the emerging science of chronobiology for any hints on how to improve their efficiency and cardiac results. Many doctors and trainers looked for correlations between time of day, exertion, and recovery in the quest for an ideal sequence. Among them, cardiac surgeon Irving Dardik, founding chair of the U.S. Olympic Sports Medicine Committee (and the first to use umbilical veins for bypass surgery), discovered that he could help runners and patients alike modulate their cardiac rhythms by having them exercise and rest at very particular intervals during the day.[25] He based his work on fairly precise but seemingly esoteric relationships between his subject's cardiac and respiratory rhythms and what he called the "superwave" of lunar, solar, and other cycles. The work got great results for many patients contending with chronic (as in *chronos*, get it?) illness, but Dardik's enthusiasm for the broader applications of his theories—including cold fusion research and the treatment of multiple sclerosis—led to widespread derision and the revocation of his license to practice medicine.

The genie was out of the bottle, though, and the notion that there were biorhythms at play in health and performance became a working principle for sports trainers and alternative healthcare practitioners. What had previously been treated as folklore about the influence of the seasons and the lunar phases on human activity was now being confirmed by new information about the brain and its changing states. New insights into brain chemistry began to validate the premise that our moods and abilities change cyclically over time, along with the wash of neurotransmitters in which our gray matter happens to be bathing at any particular moment. Psychopharmacologist and researcher Dr. David Goodman left the mainstream neuroscience community and functioned in academic exile for over thirty years to study—on subjects including himself—how the brain boosts different neurochemicals at different times of the month in order to adapt to environmental shifts over the lunar cycle. He discovered that there are four main brain states that each dominate over a single week in the twenty-eight-day lunar cycle.[26]

Meanwhile, Dr. Joel Robertson, author of the popular depression

manual *Natural Prozac*, began to study the specific influence of each
of the dominant neurotransmitters as they relate to depression.[27] He
found that certain people naturally seek to avoid depression by boost-
ing serotonin levels, resulting in calm, peaceful mind states. Others
are attracted to arousal and excitement, and tend toward behaviors
and diets that boost "up" chemicals such as norepinephrine and dopamine. (The former would opt to abuse speed and cocaine; the latter
may gravitate toward alcohol and Valium.) As significant as the exercise and behavior regimens he developed were his conclusions about
the way different neurotransmitters favor particular mental states and
activities.

Putting all this together into a comprehensive approach to the
brain over time is Dr. Mark Filippi, founder of Somaspace.org and a
sought-after consultant to Wall Street stockbrokers and professional
sports teams looking to improve their performance, as well as more
crunchy New Age types looking to make sense of the patterns in
their lives. I interviewed him at his modest Larchmont office filled
with books by everyone from general semanticist Alfred Korzybski
to scientist David Bohm. "Why does a baseball player go into a
slump?" he asks me. "Because he's out of sync with the waves in his
world."[28] That's a sports statistic I hadn't heard about before.

Filippi argues that instead of masking our incoherence on a biochemical level through psychotropic drugs like Prozac and Valium,
we need to organize our social routines with more acute awareness
of the chronobiology informing them. He helps his clients coordi-
nate their activities and exercises with natural cycles occurring all
around and within them, in an effort to increase their efficiency, reduce their stress, and generate well-being. He is doughy but fast-
talking—an avid sports fan with a small TV on in his office 24/7, as
if it were a window to the sociocultural weather. Watching a sports
game with Filippi is like listening to an entirely alternative radio
commentary. "Look at his jaw," Filippi comments of a point guard in
a basketball game on the TV. "His ankle is going to give out." Sure

enough, two plays later the point guard is on the ground clutching his shin. "It's all connected," he says, as if the structural dynamic between jaw and ankle through "tensegrity"[29] were patently obvious.

"When we speak of *tensegrity*," Filippi explains, "we mean the capacity a system has to redistribute tension and retain the same physical shape. Our manifested reality, from the infrastructure of our cells to the street grids of the towns and cities we live in, possess a tensegrity." Most basically, tensegrity is what holds it all together. Our most internalized patterns, such as breathing, moving, and relating, embody it. Without tensegrity, we no longer exist; we're just a bunch of cells. But it's also a moving target, changing its forces and rhythms along with our external activities and internal states.

Building on his predecessors' work, Filippi has been analyzing the biochemical impact of seasonal and lunar phases in order to make sense of human rhythms and determine optimal times for both therapy and particular activities. Just as there are four solar seasons with rather obvious implications (winter is better for body repair; summer is better for exertion), there are also four corresponding moon phases, sections of the day, quarters of the hour, and even stages of breath, Filippi argues. By coordinating our internal four-part, or "four phase," rhythms with those of our greater environment, we can think, work, and interact with greater coherence. Integrating the research of Dardik, Goodman, and Robinson along with his own observations, Filippi concluded that in each moon phase the brain is dominated by a different neurotransmitter. According to Filippi, the prevalence of one chemical over the others during each week of the lunar cycle optimizes certain days for certain activities.

At the beginning of the new moon, for example, one's acetylcholine rises along with the capacity to perform. Acetylcholine is traditionally associated with attention. "The mood it evokes in us is an Energizer Bunny–like pep. That vibe can be used to initiate social interactions, do chores and routines efficiently, and strive for balance in our activities."

Nearer to the full moon, an uptick in serotonin increases self-awareness, generating both high focus and high energy. Serotonin, the chemical that gets boosted by drugs like Prozac, is thought to communicate the abundance or dearth of food resources to our brain. "When under its influence we can feel euphoric, spontaneous, and yet composed and sedate. Whereas acetylcholine worked to anchor us to our physical world, serotonin buoys us to the mental realm, allowing us to experience the physical world from an embodied, more lucid vantage point. We actually benefit from solitude at this time, as when an artist finds his muse."

Over the next week, we can enjoy the benefits of increased dopamine. This chemical—responsible for the rush one gets on heroin or after performing a death-defying stunt—is responsible for reward-driven learning. "It allows us to expand our behaviors outside of our routines, decrease our intensity, and essentially blend with the energy of the moment. If acetylcholine is the ultimate memory neurotransmitter, dopamine is the ultimate experiential one. Functionally, it serves us best when we're doing social activities we enjoy." In other words, it's party week.

Finally, in the last moon phase, we are dominated by norepinephrine, an arousal chemical that regulates processes like the fight-or-flight response, anxiety, and other instinctual behaviors. "We tend to be better off doing more structural tasks that don't involve a lot of reflection. Its binary nature lets us make decisions, act on them, and then recalibrate like a GPS with a hunting rifle. The key with norepinephrine is that if it's governed well, we experience a fluid coordination of thought and action so much so that we almost fail to feel. Everything becomes second nature." So instead of letting the natural rise of fight-or-flight impulses turn us into anxious paranoids, we can exploit the state of nonemotional, almost reptilian arousal it encourages.

Further, within each day are four segments that correspond to each of these moon phases. In the new moon phase, people will be most effective during the early morning hours, while in the second phase leading up to the full moon, people do best in the afternoon.

Admittedly, this is all a tough pill for many of us to swallow, but after my interviews with Filippi, I began working in this fashion on this book. I would use the first week of the moon to organize chapters, do interviews, and talk with friends and colleagues about the ideas I was working on. In the second, more intense week, I would lock myself in my office, set to task, and get the most writing done. In the third week, I would edit what I had written, read new material, jump ahead to whatever section I felt like working on, and try out new ideas. And in the final week, I would revisit structure, comb through difficult passages, and recode the nightmare that is my website. My own experience is that my productivity went up by maybe 40 percent, and my peace of mind about the whole process of writing was utterly transformed for the better. Though certainly anecdotal as far as anyone else is concerned, the exercise convinced me to stay aware of these cycles from now on.

Digital technology can be brought back into the equation as well, promoting altogether new levels of sync. Think of it the way biofeedback works: a person is hooked up to a bunch of equipment that monitors his heart, breath, blood pressure, galvanic skin response, and so on. That information is then fed back to him as an image on a monitor; a moving line; a pulsing light, music, or sound; or an animation. The person who wants to slow his heart rate or lower his blood pressure might do so by attempting to change a red light to a blue one. The sensors are listening and the computer is processing, but the data they feed back is the articulation of inner, cyclic activity, not its repression. Seeing this previously hidden information gives us new access to natural rhythms and the ability to either adjust them or adjust ourselves and our lives to their warning signals.

More advanced deployments of these technologies attempt to synchronize those internal rhythms with those of the outside world.

Most of us today live in cities and spend most of our time indoors, where the cues that used to alert us to the changing days, moon phases, and seasons are largely hidden from us. Our minds may know it's nighttime, but our eyes and thyroids don't. Neither do most of us know what phase the moon is in, whether the tides are high, or if the honey is ready—even though some of our immune systems may have registered that ragweed is in the air and triggered an allergic response. So, some of our senses are still connected to the cycles to which we have coordinated our organ systems for the past several hundred thousand years, while others are blinded by our artificial environments or our efforts to synchronize to the wrong signals. (What does it mean to your body when a ten-ton bus zooms past? That you're in the middle of an elephant stampede?)

In an effort to resynchronize our internal rhythms with those of nature, chronobiologists have developed computer programs that both monitor the various pulses of our organs while comparing and contrasting them with daily, lunar, and seasonal cycles. Not everyone marches to exactly the same beat, which is why these approaches are highly personalized. New, controversial, but effective exercise programs from companies with names like iHeart and LifeWaves exploit these emerging findings,[30] assigning workouts during different, specific hours of the day over the course of each month.

Believe what we may about such strategies, there is now substantial evidence that time is not generic, however interchangeable our digital devices may make it seem. Yes, the DVR means we can record our programs instead of watching them as they are broadcast, but this does not mean the experience of watching our favorite HBO drama is the same on a Wednesday afternoon as it would have been on the Sunday evening before when everyone else was watching it. We needn't be slaves to the network's schedule, but neither must we submit to the notion that we are equally disposed to all activities in all time slots.

By the same token, businesses need not submit to the schedules and sequences that are external to the social and cultural rhythms defining their product cycles. Duncan Yo-Yos, for just one example, enjoy a cyclical popularity as up and down as the motion of the toy itself. The products become wildly popular every ten years or so, and then retreat into near total stagnation. The company has learned to ride this ebb and flow, emerging with TV campaigns, celebrity spokespeople, and national tournaments every time a new generation of yo-yo aficionados comes of age.

Likewise, Birkenstock shoes rise and fall in popularity along with a host of other back-to-nature products and behaviors. Instead of resisting these trend waves and ending up with unsold stock and disappointing estimates, the company has learned to recognize the signs of an impending swing in either direction. With each new wave of popularity, Birkenstock launches new lines and opens new dealerships, then pulls back when consumer appetites level off. With this strategy, the company has expanded from making just a handful of shoes to its current offering of close to five hundred different styles.

A friend of mine who worked on smart phone applications at Apple told me that Steve Jobs always thought of product development in terms of three-year cycles. Jobs simply was not interested in what was happening in the present; he only cared about the way people would be working with technology three years in the future. This is what empowered Jobs to ship the first iMacs with no slot for floppy drives, and iPhones without the ability to play the then-ubiquitous Flash movie files. He was developing products in the present that were situated to catch (and, some may argue, create) changing consumer trends instead of simply meeting his marketing department's snapshots of current consumer demand.

This is not as easy for companies with shareholders who refuse to look at anything but quarterly reports and expect year-over-year earnings to do nothing but increase. But that is not the way either people or organizations function—especially not when they are using social networking technologies that in many ways promote the coherence of their living cultures. For while digital technology can serve to disconnect us from the cycles that have traditionally orchestrated our activities, they can also serve to bring us back into sync.

The choice of how to use them remains ours.

THE SPACE BETWEEN THE TICKS

I grew up back when family vacations meant long road trips. My brother and I would fight for control of the AAA TripTik, a plastic-bound pad of little one-page maps with a hand-drawn green-felt-tip pen line over the highway we were supposed to follow. Holding the map meant getting to look at the tiny numbers indicating the mileage between each exit, and then trying to add up how far it was to the next destination or rest stop, while out the window the trees and road signs flew by.

"I'm getting carsick," the map holder would inevitably call out, handing over navigational authority to his sibling.

"Look out the window," our dad would say. "At things far away. And stop looking at that map!"

What our father, an accountant, understood from experience if not brain science was that it's hard to focus on something up close while moving at sixty-five miles per hour down a highway. This is due to what neurologists call sensory conflict. In order to locate our own positions in space, we use multiple sources of information: sight, touch, the angles of our joints, the orientation of the inner ear, and more. When we are focused on a fine motor task or tiny detail, the scale on which this occurs just doesn't jibe with the speed at which things are flying past in our peripheral vision, or the bumps and acceleration we feel in our stomachs as the car lurches forward. Are we sitting on a couch, falling out of a tree, or running through a field? The body doesn't know from automobiles. Our inner ear says we are moving, while our eyes, staring at a fixed object like a book, tell the brain we are stationary.

We are similarly disoriented by digital time, for it tends to mix and match different scales simultaneously. A date twenty years in the future has the same size box in the Google calendar as the one tomorrow or next week. A search result may contain both the most recent Tweet about a subject as well as an authoritative text representing fifty years of scholarship—both in the same list, in the same size, and virtually indistinguishable. Just like the world whizzing by out the car window, the digital world on the other side of our computer screens tends to move out of sync with the one in which our bodies reside.

In both cases, the driver is the only one truly safe from nausea. For while the passengers—or in the digital realm, users—cannot anticipate the coming bumps or steer into curves, drivers can. For those of us contending with digiphrenia, becoming a driver means taking charge of choice making, especially when that means refusing to make any choice at all.

Digital technology is all about choices. Closer to a computer game than a continuous narrative, the digital path is no longer inevitable, but a branching hierarchy of decision points. The digital timeline moves not from moment to moment, but from choice to choice, hanging absolutely still on each command line—like the number on a digital clock—until the next choice is made and a new reality flips into place.

This freedom to choose and make choices is the underlying promise of the digital era, or of any new technology. Electric lighting gives us the freedom to choose when to sleep; asphalt gives us the choice where to drive our cars; Prozac gives us the freedom to choose an otherwise depressing lifestyle. But making choices is also inherently polarizing and dualist. It means we prefer one thing over another and want to change things to suit our sense of how things ought to be.

Our leading consumer-technology brand, Apple, makes this all too clear: using these devices is akin to taking a bite of the forbidden fruit, exchanging the ignorant holism of Eden for the self-aware choice making of adulthood. The "Tree of the Knowledge of Good and Evil," as the myth calls it, introduced humanity to the binary universe of active choice that computers now amplify for us today. As a downside, the new freedom of choice created self-consciousness and shame. Adam and Eve became self-aware and ashamed of their nudity. They were banished from the holism of Eden and went out into the world of yes and no, this and that, Cain and Abel, and good and evil.

The same is true for us today. Our digital technologies empower us to make so many choices about so many things. But the staccato nature of digital choice also thwarts our efforts to stay fully connected to our greater throughlines and to one another. Every choice potentially brings us out of immersive participation and into another decision matrix. I am with my daughter, but the phone is vibrating with a new instant message. Even if I choose to ignore the message and be with her, I have been yanked from the intimate moment by the very need to make a choice. Of course, I can also choose to turn off the phone—which involves pulling it out of my pocket and changing a setting—or just leave it and hope it doesn't happen again.

For many of us trying to reconcile our real and virtual scales of existence, there is almost a feeling of operating at different speeds or in multiple time zones. This sensation comes from having to make precise, up-close, moment-to-moment choices while simultaneously attempting to experience the greater flow we associate with creativity and productivity. Like when you're writing with great flow and energy, but your Microsoft Word program puts a little green line under one word. Do you stop and check, or do you keep going? Or both? Just like the kid in the backseat of the car, we get a kind of vertigo.

The ancient Greeks would probably tell us our troubles stem from our inability to distinguish between the two main kinds of time, *chronos* and *kairos*. It's as if they understood that time is simply too multifaceted to be described with a single word. *Chronos* is the kind of time that's registered by the clock: chronology. It's not time itself but a particular way of understanding time by the clock. That's what we literally mean when we say "three o'clock." This is time *of the clock*, meaning belonging to the clock, or *chronos*.

Kairos is a more slippery concept. Most simply, it means the right or opportune moment. Where *chronos* measures time quantitatively, *kairos* is more qualitative. It is usually understood as a window of opportunity created by circumstances, God, or fate. It is the ideal time to strike, to propose marriage, or to take any particular course of action. *Carpe diem*. *Kairos* is perfect timing relative to what's going on, where *chronos* is the numerical description of what happens to be on the clock right then. *Chronos* can be represented by a number; *kairos* must be experienced and interpreted by a human.

While clocks may have suggested that we live in a world bound by *chronos*, digitality asks us to embody *chronos* itself. Where the arms of the clock passed through the undefined, unmeasured spaces between numbers, digital technology registers only *chronos*. It does not exist between its pulses. This is why we call it an asynchronous technology: it does not pass. It is the ticks of the clock but none of the space between. Each new tick is a new line of a code, a new decision point, another division—all oblivious to what happens out there in the world—except when it gets a new discrete input from one of us.

Digital time ignores nearly every feature of *kairos*, but in doing so may offer us the opportunity to recognize *kairos* by its very absence. Clocks initially disconnected us from organic time by creating a metaphor to replace it. Digital time is one step further removed, replacing what it was we meant by "time" altogether. It's a progression akin to what postmodern philosopher Jean Baudrillard called the "precession of the simulacra." There is the real world, then there are the metaphors and maps we use to represent that world, and then there is yet another level of activity that can occur on those maps—utterly disconnected from the original. This happens because we have grown to treat the maps and symbols we have created as if they were the underlying reality. Likewise, we started with this amorphous experience of rhythms that we called time. We created the analog clock to represent the aspects of time we could represent with a technology. Then, with digital readouts, we created a way of representing what was happening on that clock face. It is twice removed from the original.

Now that *chronos* has been fully freed from the cycles and flows through which we humans experience time, we can more easily differentiate between the kinds of clocks and time we are using. We can stop forcing our minds and bodies to keep up with digital *chronos* while also ceasing to misapply our digital technologies to human processes. We come to fully recognize the difference between *chronos* and *kairos*, or between time and timing.

Or think of it this way: Digital technology is more like a still-life picture. A sample. It is frozen in time. Sound, on the other hand, is audible only over time. We hear sound as it decays. Image may be thought of as *chronos*, where sound is more like *kairos*. Not surprisingly, the digital universe is a visual one: people staring silently at screens, where the only sounds in the room are the keys and mouse clicks.

Our analog technologies anchored us temporally in ways our digital ones don't. In a book or a scroll, the past is on our left, and the future is on our right. We know where we are in linear time by our position in the paper. While the book with its discrete pages is a bit more sequential than the scroll, both are entirely more oriented in time than the computer screen. Whichever program's window we may have open at the moment is the digital version of now, without context or placement in the timeline. The future on a blog is not to one side, but above—in the as-yet-unposted potential. The past isn't to the other side, but down, in and among older posts. Or over there, at the next hypertext link. What is next does not unfold over time, but is selected as part of a sequence.

In this context, digiphrenia comes from confusing *chronos* with *kairos*. It happens when we accept the digital premise that every moment must potentially consist of a decision point or a new branch. We live perched atop the static points of *chronos*, suffering from the vertigo of no temporal context. It's akin to the discomfort many LP listeners had to early CD music and low-resolution digital music files, whose sample rates seemed almost perceptible as a staccato sawtooth wave buzzing under the music. It just didn't feel like continuous sound and didn't have the same impact on the body.

Digital audio is markedly superior in many respects. There is no "noise"—only signal. There are no "pops" on the record's surface, no background hiss. But without those references, our experience becomes frictionless—almost like the vertigo we experience when zooming in or out of a Google map too rapidly. There's no sense of scale—no background to the foreground. None of the cues that create a sense of organic connection to the medium. There is no messy handwriting to be deciphered, only ACII text. Every copy is the original. And it is perfect—at least when it's there. When it is not, as in the case of the spaces between the samples in a digital audio recording, there is nothing. For many kinds of activities, this doesn't matter at all. The fixed type of the printing press already removed pretty much every sign of analog personality from text. For human purposes, there is no real difference between the text in a digital book and a physical one; the only difference is the form factor of the book versus that of the tablet. That's because a symbol system like text is already abstracted and just as well represented digitally.

When there is a direct communication with the senses, on the other hand, the difference becomes a lot clearer. Like a fluorescent lightbulb, which will perceptibly flicker at 60 hertz along with the alternating current of the house, digital technologies are almost perceptibly on/off. They create an environment, regardless of the content they are expressing. This is what Marshall McLuhan meant by "the medium is the message." A lightbulb creates an environment, even though it has no content. Even without a slide or movie through which to project an image onto the wall, the light itself creates an environment where things can happen that otherwise wouldn't. It is an environment of light.

With digital technology, the environment created is one of choice. We hop from choice to choice with no present at all. Our availability to experience flow or to seize the propitious moment is minimized as our choices per second are multiplied by a dance partner who doesn't see or feel us. Our rhythms are dictated by the pulses of required inputs and incoming data. It is not a stream but a series of points along a line. Yes, we have the ability to make more choices, but in the process we become primarily choosers. The *obligation* to choose—to "submit" as the button compels us—is no choice at all. Especially when it prevents us from achieving our own sense of flow and rhythm.

The first experience most of us had of this sort of forced choice was call waiting—the interruptive beep letting us know we had the option of putting our current conversation on hold and responding to whoever was now calling us. By utilizing the feature and checking on the incoming call, we are introducing a new choice into our flow. In the process, we are also saying, "Hold on while I check to make sure there's not someone calling whom I want to speak with more than I want to speak with you." Sure, we can justify that it might be an emergency, but it's really just a new decision point.

Now with caller ID in addition to our call waiting, we can visually check on the source of the incoming call without our current caller even knowing. We pull the phone away from our ear, imperceptibly disconnecting from the call for long enough to scan the incoming number. Then we return to the conversation in process, only half listening while we inwardly debate whether to interrupt the conversation with the now certain fact that whatever the person on the other end is saying is less important to us than the alternative, the incoming caller. (It was easier when everyone heard the beep, and the insult seemed less willful.)

The real crime here is not simply the indignity suffered by our rejected talk partner, but the way we so easily allow the sanctity of our moment to be undermined. It assumes that *kairos* has no value—that if there is a moment of opportunity to be seized, that moment will break into our flow from the outside, like a pop-up ad on the Web. We lose the ability to imagine opportunities emerging and excitement arising from pursuing whatever we are currently doing, as we compulsively anticipate the next decision point.

Clay Shirky correctly distinguishes this problem from the overused term "information overload," preferring instead to call it "filter failure." In a scarce mediaspace dominated by books, printing a text meant taking a financial risk. The amount of information out there was limited by the amount of money publishers and advertisers were willing to spend. Now that information can be generated and distributed essentially for free, the onus is upon the receiver to filter out or even stop the incoming traffic. Even though each one seems to offer more choices and opportunities, its very presence demands a response, ultimately reducing our autonomy.

Once we make the leap toward valuing the experience of the now and the possibilities of *kairos*, we can begin to apply some simple filters and mechanisms to defend it. We can set up any cell phone to ring or interrupt calls only for family members; we can configure our computers to alert us to only the incoming emails of particular people; and we can turn off all the extraneous alerts from everything we subscribe to. Unless we want our entire day guided by the remote possibility that a plane may crash into our office building, we need to trust that we can safely proceed on the assumption that it won't.

While there is tremendous value in group thinking, shared platforms, and networked collaboration, there is also value in a single mind contemplating a problem. We can defend our access to our personal *kairos* by letting the digital care for the *chronos*. Email lives outside time and can sit in the inbox until we are ready to read it. This should not be guilt provoking. The sender of the email is the one who relegated this missive to the timeless universe. Once sent, it is no longer part of our living, breathing cycling world of *kairos* but of the sequential realm of *chronos*. Email will form a stack, just like the stacks of processes in a computer program, and wait until we open it.

When I visit companies looking to improve their digital practices, I often suggest office workers attempt not to answer or check on email for an entire hour—or even two hours in a row. They usually reel in shock. Why is this so hard for so many of us? It's not because we need the email for our productivity, but because we are addicted to the possibility that there's a great tidbit in there somewhere. Like compulsive gamblers at a slot machine rewarded with a few quarters every dozen tries, we are trained to keep opening emails in the hope of a little shot of serotonin—a pleasant ping from the world of *chronos*. We must retrain ourselves instead to see the reward in the amount of time we get to spend in the reverie of solo contemplation or live engagement with another human being. Whatever is vibrating on the iPhone just isn't as valuable as the eye contact you are making right now.

A friend of mine makes her living selling homemade candles through the craft site etsy.com. As her business got more successful, more orders would come through. By habit, she stopped whatever she was doing and checked her email every time a new message dinged through. If it was an order, she opened it, printed it out, and filled it before returning to her work melting wax, mixing scents, and dipping her wicks. Her routine became broken up and entirely un-fun, until it occurred to her to let the emails stack up all day, and then process them all at once. She used an automatic reply to all incoming orders, giving customers the instant feedback they have come to expect, but kept her flesh-and-blood candle maker self insulated from the staccato flow of orders. At 3 p.m. each day, her reward was to go to the computer and see how many orders had come in since morning. She had enough time to pack them all at once while listening to music, and then take them to the post office before closing. She achieved greater efficiency while also granting herself greater flow.

The digital can be stacked; the human gets to live in real time. This experience is what makes us creative, intelligent, and capable of learning. As science and innovation writer Steven Johnson has shown, great ideas don't really come out of sudden eureka moments, but after long, steady slogs through problems.[31] They are slow, iterative processes. Great ideas, as Johnson explained it to a TED audience, "fade into view over long periods of time." For instance, Charles Darwin described his discovery of evolution as a eureka moment that occurred while he was reading Malthus on a particular night in October of 1838. But Darwin's notebooks reveal that he had the entire theory of evolution long before this moment; he simply hadn't fully articulated it yet.

As Johnson argues it, "If you go back and look at the historical record, it turns out that a lot of important ideas have very long incubation periods. I call this the 'slow hunch.' We've heard a lot recently about hunch and instinct and blink-like sudden moments of clarity, but in fact, a lot of great ideas linger on, sometimes for decades, in the back of people's minds. They have a feeling that there's an interesting problem, but they don't quite have the tools yet to discover them." Solving the problem means being in the right place at the right time—available to the propitious moment, the *kairos*.

Perhaps counterintuitively, protecting what is left of this flow from the pressing obligation of new choices gives us a leg up on innovation.

Extracting digital processes from our organic flow not only creates the space for opportune timing to occur, it also helps prevent us from making inopportune gaffes. Like the famous television sketch where Lucy frantically boxes chocolates on the assembly line, we attempt to answer messages and perform tasks at the rate they come at us. And like Lucy, we end up jamming the wrong things in the wrong places. Gmail gives us a few minutes to recall messages we may have sent in error, but this doesn't stop us from replying hastily to messages that deserve our time, and creating more noise than signal.

Comments sections are filled with responses from people who type faster than they think and who post something simply because they know they will probably never have time to find the discussion again. The possibility that someone might actually link to or re-Tweet a comment leads people to make still more, turning posting into just as much a compulsion as opening email messages. As our respected leaders post their most inane random thoughts to our Twitter streams, we begin to wonder why they have nothing better to do with their time, or with ours. When an actor with the pop culture simplicity of Ashton Kutcher begins to realize that his unthinking posts can reflect negatively on his public image,[32] it should give the rest of us pause. But we would need to inhabit *kairos* for at least a moment or two in order to do that.

The result is a mess in which we race to make our world conform to the forced yes-or-no choices of the digiphrenic.

I once received an email from a college student in Tennessee who had been recruited by a political group to protest against leftist professors. Since I was scheduled to speak at the university in a month, she had studied my website for over ten minutes, trying to figure out if I was a leftist. After perusing several of my articles, she was still unable to determine exactly which side of the political spectrum I was on. Could I just make this easier for her and tell her whether or not I was a leftist, so that she knew whether to protest my upcoming speech? I pointed her to some of my writing on economics and explained that the Left/Right categorization may be a bit overdetermined in my case. She thanked me but asked me to please just give her a yes-or-no answer.

"Yes and no," I replied, breaking the binary conventions of digital choice making. I didn't hear back.

DO DRONE PILOTS DREAM OF ELECTRIC KILLS?

I was working on a story about Predator drones, the unmanned aircraft that the US Air Force flies over war zones in the Middle East and Central Asia collecting reconnaissance, taking out targets, and occasionally launching a few Hellfire missiles. The operators—pilots, as they're called—sit behind computer terminals outfitted with joysticks, monitors, interactive maps, and other cockpit gear, remotely controlling drones on the other side of the world.

I was most interested in the ethical challenges posed by remote control warfare. Air Force commanders told me that within just a few years a majority if not all American fighter planes would be flown remotely by pilots who were thousands of miles away from the war zone. The simulation of flight, the resolution of the cameras, the instantaneousness of feedback, and the accuracy of the controls have rendered the pilot working through virtual reality just as effective as if he or she were in the cockpit. So why risk human troops along with the hardware? There's no good reason, other than the nagging sense that on some level it's not fair. What does it mean to fight a war where only one side's troops are in jeopardy, and the other side may as well be playing a video game? Will our troops and our public become even more disconnected from the human consequences and collateral damage of our actions?

To my great surprise, I found out that the levels of clinical distress in drone crews were as high, and sometimes even higher, than those of crews flying in real planes.[33] These were not desensitized video-game players wantonly dropping ordinance on digital-screen pixels that may as well have been bugs. They were soul-searching, confused, guilt-ridden young men, painfully aware of the lives they were taking. Thirty-four percent experienced burnout, and over 25 percent exhibited clinical levels of distress. These responses occurred in spite of the Air Force's efforts to select the most well adjusted pilots for the job.

Air Force researchers blamed the high stress levels on the fact that the drone pilots all had combat experience and that the drone missions probably caused them to re-experience the stress of their real-world missions. After observing the way these pilots work and live, however, I'm not so sure it's their prior combat experience that makes drone missions so emotionally taxing, as it is these young men's *concurrent* life experience. Combat is extraordinarily stressful, but at least battlefield combat pilots have one another when they get out of their planes. They are far from home, living the war 24/7 from their military base or aircraft carrier. By contrast, after the drone pilot finishes his mission, he gets in his car and drives home to his family in suburban Las Vegas. He passes the mashed potatoes to his wife and tries to talk about the challenges of elementary school with his second grader, while the video images of the Afghan targets he neutralized that afternoon still dance on his retinas.

In one respect, this is good news. It means that we can remain emotionally connected to the effects of our virtual activities. We can achieve a sort of sync with things that are very far away. In fact, the stress experienced by drone pilots was highest when they either killed or witnessed the killing of people they had observed over long periods of time. Even if they were blowing up a notorious terrorist, simply having spied on the person going through his daily routine, day after day, generated a kind of sympathetic response—a version of sync.

The stress, depression, and anxiety experienced by these soldiers, however, came from living two lives at once: the life of a soldier making kills by day and the one of a daddy hugging his toddler at night. Technology allows for this dual life, this ability to live in two different places—as two very different people—at the same time. The inability to reconcile these two identities and activities results in digiphrenia.

Drone pilots offer a stark example of the same present shock most of us experience to a lesser degree as we try to negotiate the contrast between the multiple identities and activities our digital technologies demand of us. Our computers have no problem functioning like this. When a computer gets a problem, or a series of problems, it allocates a portion of its memory to each part. More accurately, it breaks down all the tasks it needs to accomplish into buckets of similar tasks, and then allocates a portion of its memory—its processing resources—to each bucket. The different portions of memory then report back with their answers, and the chip puts them back together or outputs them or does whatever it needs to next.

People do not work like this. Yes, we do line up our tasks in a similar fashion. A great waiter may scan the dining room in order to strategize the most efficient way to serve everyone. So, instead of walking from the kitchen to the floor four separate times, he will take an order from one table, check on another's meal while removing a plate, and then clear the dessert from a third table and return to the kitchen with the order and the dirty dishes. The human waiter strategizes a linear sequence.

The computer chip would break down the tasks differently, lifting the plates from both tables simultaneously with one part of its memory, taking the order from each person with another part (broken down into as many simultaneous order-taking sections as there are people), and check on the meal with another. The human figures out the best order to do things one after the other, while the chip divides itself into separate waiters ideally suited for each separate task. The mistake so many of us make with digital technology is to imitate rather than simply exploit its multitasking capabilities. We try to maximize our efficiency by distributing our resources instead of lining them up. Because we can't actually be more than one person at the same time, we experience digiphrenia instead of sync.

The first place we feel this is in our capacity to think and perform effectively. There have been more than enough studies done and books written about distraction and multitasking for us to accept—however begrudgingly—the basic fact that human beings cannot do more than one thing at a time.[34] As Stanford cognitive scientist Clifford Nass has shown pretty conclusively, even the smartest university students who believe they are terrific at multitasking actually perform much worse than when they do one thing at a time. Their subjective experience is that they got more done even when they accomplished much less, and with much less accuracy. Other studies show that multitasking and interruptions hurt our ability to remember.

We do have the ability to do more than one thing at a time. For instance, we each have parts of our brain that deal with automatic functions like breathing and beating while our conscious attention focuses on a task like reading or writing. But the kind of plate spinning we associate with multitasking doesn't really happen. We can't be on the phone while watching TV; rather, we can hold the phone to our ear while our eyes look at the TV set and then switch our awareness back and forth between the two activities. This allows us to enjoy the many multifaceted, multisensory pleasures of life—from listening to a baseball game while washing the car to sitting in the tub while enjoying a book. In either case, though, we stop focusing on washing the car in order to hear about the grand slam and pause reading in order to decide whether to make the bath water hotter.

It's much more difficult, and counterproductive, to attempt to engage in two active tasks at once. We cannot write a letter while reconciling the checkbook or—as the rising accident toll indicates—drive while sending text messages. Yet the more we use the Internet to conduct our work and lives, the more compelled we are to adopt its processors' underlying strategy. The more choices are on offer, the more windows remain open, and the more options lie waiting. Each open program is another mouth for our attention to feed.

This competition for our attention is fierce. Back in the mid-1990s, *Wired* magazine announced to the world that although digital real estate was infinite, human attention was finite; there are only so many "eyeball hours" in a day per human. This meant that the new market—the new scarcity—would be over human attention itself. Sticky websites were designed to keep eyeballs glued to particular spots on the Internet, while compelling sounds were composed to draw us to check on incoming-message traffic. In a world where attention is the new commodity, it is no surprise that diagnoses of the formerly obscure attention deficit disorder are now so commonplace as to be conferred by school guidance counselors. Since that *Wired* cover in 1997, Ritalin prescriptions have gone up tenfold.

Kids aren't the only ones drugging up. College students and younger professionals now use Ritalin and another form of speed, Adderall, as "cognitive enhancers."[35] Just as professional athletes may use steroids to boost their performance, stockbrokers and finals takers can gain an edge over their competitors and move higher up on the curve. More than just keeping a person awake, these drugs are cognitive accelerators; in a sense, they speed up the pace at which a person can move between the windows. They push on the gas pedal of the mind, creating momentum to carry a person through from task to task, barreling over what may seem like gaps or discontinuities at a slower pace.

The deliberate style of cognition we normally associate with reading or contemplation gives way to the more superficial, rapid-fire, and compulsive activities of the net. If we get good enough at this, we may even become what James G. March calls a "fast learner," capable of grasping the gist of ideas and processes almost instantaneously. The downside is that "fast learners tend to track noisy signals too closely and to confuse themselves by making changes before the effects of previous actions are clear."[36] It's an approach that works better for bouncing from a Twitter stream to a blog comments field in order to parse the latest comments from a celebrity on his way to rehab than it does for us to try to solve a genetic algorithm.

But it's also the quiz-show approach now favored by public schools, whose classrooms reward the first hand up. Intelligence is equated with speed, and accomplishment with the volume of work finished. The elementary school where I live puts a leaf on the wall for every book a child reads, yet has no way of measuring or rewarding the depth of understanding or thought that took place—or didn't. More, faster, is better. Kids compete with the clock when they take their tests, as if preparing for a workplace in which their boss will tell them "pencils down." The test results, in turn, are used to determine school funding and teacher salaries. All children left behind.

The more we function on the level of fast learning, the less we are even attracted to working in the other way. It's a bit like eating steak and potatoes after you've had your chocolate. And the market—always acting on us but especially so in the artificial realm of the Internet—benefits in the short term from our remaining in this state, clicking on more things, signing into more accounts, and consuming more bytes. Every second we spend with just one thing running or happening is a dozen market opportunities lost.

The digital realm, left to its own devices or those of the marketplace, would atomize us all into separate consuming individuals, each with our own iPhone and iTunes account, downloading our own streams of media. Our connections to one another remain valuable only to the extent they can be mediated. If we are kissing someone on the lips, then we are not poking anyone through our Facebook accounts.

Of course, many of us try to do both. I have a friend, a psychotherapist, who often conducts sessions with patients over the phone when circumstances prevent a live meeting. What concerns me most about these sessions is the freedom he feels to send me text messages during those sessions, telling me to check out something he just noticed online! After I recover from the distraction from my own workflow, I start to feel depressed.

When the social now is relegated to the multitasking digital environment, we may expect the results we have been witnessing: teen suicides, depression, higher stress, and a greater sense of disconnection. It's not because digital technology is inherently depressing, but, again, it's because we are living multiple roles simultaneously, without the time and cues we normally get to move from one to the other.

In the real world, 94 percent of our communication occurs nonverbally.[37] Our gestures, tone of voice, facial expressions, and even the size of our irises at any given moment tell the other person much more than our words do. These are the cues we use to gauge whether someone is listening to us, agrees with us, is attracted to us, or wants us to shut up. When a person's head nods and his irises dilate, we know—even just subconsciously—that he agrees with us. This activates the mirror neurons in our brains, feeding us a bit of positive reinforcement, releasing a bit of dopamine, and leading us further down that line of thought.

Without such organic cues, we try to rely on the re-Tweets and likes we get—even though we have not evolved over hundreds of millennia to respond to those symbols the same way. So, again, we are subjected to the cognitive dissonance between what we are being told and what we are feeling. It just doesn't register in the same way. We fall out of sync.

We cannot orchestrate human activity the same way a chip relegates tasks to the nether regions of its memory. We are not intellectually or emotionally equipped for it, and altering ourselves to become so simply undermines the contemplation and connection of which we humans are uniquely capable. With this knowledge, however, we can carefully and consciously employ digital-era sync to the processes in our lives and businesses for which digiphrenia is not a liability.

Real-time supply-chain management, for example, such as the system employed by retail clothing chain Zara, syncs the store checkout with the production line. At the same moment the scanner at the cash register identifies the yellow T-shirt, the information is sent to the production facility, the facility's suppliers, and so on, all the way down the line. As one leading logistics company explains, "Synchronization of demand/supply information minimizes work-in-process and finished goods inventories up and down the channel, dampens the 'bullwhip effect' as products are pulled through the distribution pipeline, reduces costs overall, and matches customer requirements with available products."[38]

This rapid production process reduces the time from design to delivery to just a couple of weeks, making Zara famous in Spain for being able to knock off fashions from the Paris runways and get them into their store windows before they even appear in *Vogue*. Subjecting designers to this schedule or forcing them to submit to the metrics of cash register feedback would no doubt induce digiphrenia or worse. (Yes, many youth-fashion companies do just that.) But for an industry that simply copies, manufacturers, and distributes, such sync is the goal. The feedback from people is used to program the machines and not the other way around.

The same goes for the trucking industry, where computers can track the location, contents, and available capacity of thousands of trucks at once, adjusting routes and maximizing the efficiency of the entire system. Stress on the human trucker goes down, not up, as he manages to make an extra haul instead of simply riding all the way with an empty trailer after a drop.

On the Web, with the emergence of HTML 5, we are finally witnessing the disappearance of the split between ad and store. In the upcoming rather holistic model of online marketing, the banner ad is no longer a click-through to some other website, but it is the store itself. The ad unfolds and the customer can make a purchase right there and then.* You will be able to buy my next book wherever it is online that you're finding out about it, without going to Amazon. By embedding stores directly into ads, we reduce the digiphrenia and increase the sales.

Any company today can differentiate itself by imposing less digiphrenia on its employees and customers. A robo-answering computer may save your company the cost of a human receptionist, while instead costing your customers and suppliers hundreds of man-hours in menu navigation. Except in cases where you are actually trying to avoid this communication (such as health insurance appeals desks or cell phone cancellation specialists who succeed when we give up), it makes more sense for you to stop externalizing the cost to your clientele. It's the smallest businesses that should be using automatic answering machines, since they can least afford to be interrupted.

In all these cases, our ability to experience sync over digiphrenia can be traced to the extent to which we are the programmers of our own and our businesses' digital processes. In the digital realm we are either the programmers or the programmed—the drivers or the passengers.

My dad wound up his own alarm clock every night, and I always thought the reason he wasn't so alarmed by the chiming of the

---

* The discontinuous experience of leaving where you are online in order to go buy something somewhere else was never real, anyway. Websites don't exist in particular places, and we don't move anywhere as we click from one to the other.

bell as I was by the buzzing of my digital clock had less to do with the different sounds than the fact that he had wound the bell himself. His relationship to the ringing was different, because it was an expression of his own muscle, his own kinetic energy. In a sense, he was the source of the bell, where my buzz was fueled by some power plant, somewhere else. No wonder it always went off before I was ready to wake up.

OVERWINDING

THE SHORT FOREVER

Too many of my favorite basketball players got injured in 2012. Just two months into the season, a significantly higher number of players were on the bench with injuries than the year before. Previously, there were on average 7.3 injuries to basketball players each day. In 2012 there were nearly 10 new injuries every day.[1]

Basketball fans knew to blame this on the shortened and condensed basketball schedule resulting from an extended lockout in the fall. By the time players and team owners negotiated a contract, several months of the playing season had passed, so a compressed schedule, with more games per week, was thrown together. As sports publications unanimously agreed, with virtually no training camp or preseason ramp-up, players were on the court and getting more hurt than usual. Or so it seemed.

Even though the numbers appeared to show that the shorter, packed schedule had led to more stress and injury, our perception of what actually happened was being distorted by time. Yes, there were more injuries per day in the NBA, but that's because there were more total games being played per day. The number of injuries *per game* actually remained constant over the past three seasons. The real difference between seasons is that it took fewer days on the calendar to accumulate fifty games' worth of injuries. We watched them occurring faster, so everyone—even the experts—assumed there were more. In hindsight, we know that the increased injury rate was an illusion. It was the same sort of illusion that makes us believe that shark attacks are on the rise simply because we have a global media capable of reporting them wherever and whenever they happen— which they cannot help but do.

What was real, however, is that players experienced more injuries in less time. This meant less time for recovery between injuries, more games played while wounds were still healing, and the anxiety of sitting out more games for every week spent on the bench. So the arithmetically disproved fans and sportswriters were still right about something, even though they didn't have the accurate numbers to prove that on some very real level things were worse. The long-term effects of this compressed schedule of stress, injury, and recovery have yet to be determined.

It's a bit like what a figure skater in an ice show experiences when stuck in the outermost position in one of those giant group pinwheel maneuvers. The skater on the outside is traveling the same number of revolutions as the skaters closer to the center, even though the outside skater is moving much faster and covering much more distance. They're all in the very same circle but functioning on very different scales of activity. We understand this viscerally as we watch that skater on the very end of the line race desperately around the rink while the ones in the middle smile and barely move. But we tend to lose sight of these different scales of time as they play out in the world around us.

Scientist Freeman Dyson attempted to clarify this for us when he came up with the concept of temporal diversity. As he came to realize, the survival of a species depends on adaptation and learning on six distinct timescales. On the shortest, most immediate scale, species must exist from year to year. The unit of survival for this year-to-year existence is the individual life form. Over decades, the unit of survival is the family, whose multiple generations last much longer than any single individual's life span. Over centuries, it's the tribe or nation. Over millennia, it is an entire culture. Over tens of millennia, it is the species itself, slowly evolving or surrendering to an evolved competitor. And over eons, the unit of survival is "the whole web of life of our planet."[2] Human beings have endured as a result of our ability to adapt on all six scales of being and to balance the conflicting demands of each.

This notion of temporal diversity offers a new way of under-standing the particular characteristics of different timescales. While chronobiologists looked at the various natural cycles influencing the processes of life, proponents of temporal diversity are encouraging us to understand and distinguish between the different rates at which things on different levels of existence change.

Former Merry Prankster and *Whole Earth Catalog* founder Stew-art Brand applied temporal diversity to different levels of society. In his book *The Clock of the Long Now*, he argues that we live in a world with multiple timescales, all moving simultaneously but at differ-ent speeds. Brand calls it the order of civilization. Nature, or geo-logical time, moves the slowest—like the skater in the middle of the pinwheel. This is the rate at which glaciers carve out canyons or species evolve gills and wings—over eons. On the next level is cul-ture, such as that of the Chinese or the Jews—which lasts millen-nia. On the next concentric ring comes governance—the rather long-lasting systems of monarchies and republics. The next level is infrastructure—the roads and utilities those governments build and rebuild. Faster yet is commerce that occurs through that infrastructure. And finally the outermost ring is that of fashion—the ever-changing styles and whims that keep the wheels of commerce fed.

Where we get into trouble, however, is when we lose the ability to distinguish between the different scales of time and begin to subject one level of activity to the time constraints of another. For basketball players, it was subjecting their game schedule—a culture of sorts—to the more temporally abstract timing of contracts and union negotiations. For politicians, it is the attempt to act on the timescale of government while responding to polls conducted on the scale (and value system) of fashion. For companies, it is trying to create value on the timescale of infrastructure, while needing to meet the investment requirements on the timescale of commerce. For the environment, it is fixing our focus on where to find the cheapest gallon of gasoline, with little awareness of the hundreds of thousands of years it took for the energy within it to be accumulated and compressed. Or throwing a plastic bottle into the trash because the recycle bin is thirty seconds out of our way—even though the plastic will become part of the planet's landfill for hundreds of years until it decomposes into toxic chemicals for hundreds or thousands more.

For Brand, the solution is to expand our awareness of the larger, slower cycles. He is working with inventor Danny Hillis to build a 10,000-year clock—a clock of the "long now" that changes our orientation to time. His hope is that by beholding this tremendous time-keeping structure in the desert (itself the product of a multigenerational effort), we will be able to experience or at least perceive the bigger cycles that evade us in our daily schedules. In addition, instead of writing our years with four digits, Brand encourages us to use five, as in 02020 instead of 2020, keeping us aware of the much larger timescales on which important activity is still occurring. We could still operate on the timescale of the seasonally changing fashions or TV schedules while remaining cognizant of the greater cycles binding humanity to the cosmos.

Maintaining such a multiple awareness may be a worthy goal, but it's also a treacherous path—particularly for those of us living in the new eternal present. Dyson and Brand both understand the way all activities, particularly in a Western culture, tend toward the faster layers of pacing on the periphery of the pinwheel. When time is money, everything tends to occur in the timescale of fashion. As Brand puts it, "There is always an alternative to the present urgency—and it's not a vacation, it's acknowledging deeper responsibility."[3] Still, the only ones who seem to get to live with awareness of the larger cycles are either indigenous people herding sheep in the mountains or climate scientists measuring our progress toward environmental catastrophe. What happens when those of us living at the pace of fashion try to insert an awareness of these much larger cycles into our everyday activity?

In other words, what's it like to envision the ten-thousand-year impact of tossing that plastic bottle into the trash bin, all in the single second it takes to actually toss it? Or the ten-thousand-year history of the fossil fuel being burned to drive to work or iron a shirt? It may be environmentally progressive, but it's not altogether pleasant. Unless we're living in utter harmony with nature, thinking in ten-thousand-year spans is an invitation to a nightmarish obsession. It's a potentially burdensome, even paralyzing, state of mind. Each present action becomes a black hole of possibilities and unintended consequences. We must walk through life as if we had traveled into the past, aware that any change we make—even moving an ashtray two inches to the left—could ripple through time and alter the course of history. It's less of a Long Now than a Short Forever.

This weight on every action—this highly leveraged sense of the moment—hints at another form of present shock that is operating in more ways and places than we may suspect. We'll call this temporal compression *overwinding*—the effort to squish really big timescales into much smaller or nonexistent ones. It's the effort to make the "now" responsible for the sorts of effects that actually take real time to occur—just like overwinding a watch in the hope that it will gather up more potential energy and run longer than it can.

Overwinding happens when hedge funds destroy companies by attempting to leverage derivatives against otherwise productive long-term assets. Yet—as we'll see—it's also true when we try to use interest-bearing, long-distance central currencies to revive real-time local economies. A currency designed for long-term storage and investment doesn't do so well at encouraging transactions and exchange in the moment. Overwinding also happens when we try to experience the satisfying catharsis of a well-crafted five-act play in the random flash of a reality show. It happens when a weightlifter takes steroids to maximize the efficiency of his workouts by growing his muscles overtime. In these cases it's the *over*-winding that leads to stress, mania, depletion, and, ultimately, failure.

But in the best of circumstances, this same temporal compression can be applied deliberately and with an awareness of the way different timescales interact and support one another. Instead of overwinding time, we spring-load it so it's available to us when we need it. It's what happens when Navy SEALs prepare for months or years for a single type of maneuver, compressing hundreds of hours of training into a mission that may unwind in seconds when they are called upon to free a hostage or to take out an enemy. It's studying for a test over weeks (instead of cramming overnight), so that the things we've learned stay in our memory even after the one-hour exam is over. It's the science of storing wind energy in batteries, or of letting government work on problems that won't produce returns quickly enough for the marketplace.

Those of us who can't navigate these different temporal planes end up bingeing and purging in a temporal bulimia. To distinguish between spring-loading and overwinding we must learn to recognize the different timescales on which activity occurs and how to exploit the leverage between them without getting obese or emaciated, paralyzed or burned out.

TIME BINDING

Alfred Korzybski, the founder of the Institute of General Semantics, was fascinated by the way human beings—unlike any other life-form—could exploit temporal compression, or what he called time binding. As Korzybski saw it, plants can bind energy: Through photosynthesis, they are able to take energy from the sun and store it in chemical form. Days of solar energy are bound, or compressed, into the cells of the plant. One step up from that are animals that bind space. Because they can move around, they can gather food, avoid dangers, and exploit the energy and resources of an area much larger than that of a rooted plant. The squirrel binds, or compresses, the energy of a whole acre's worth of nuts and seeds, even though his body is only a foot long.

Human beings, however, can bind time. We can take the experiences of one generation and pass it on to the next generation through language and symbols. We can still teach our children things like hunting or fishing in real time, but our lessons can also be compressed into stories, instructions, and diagrams. The information acquired by one generation can be passed on more efficiently than if each subsequent generation needed to learn everything through experience. Each new generation can begin where the former generation left off. In Korzybski's words:

> In the human class of life, we find a new factor, non-existent in any other form of life; namely, that we have a capacity to collect all known experiences of different individuals. Such a capacity

increases enormously the number of observations a single individual can handle, and so our acquaintance with the world around, and in, us becomes much more refined and exact. This capacity, which I call the time-binding capacity, is only possible because, in distinction from the animals, we have evolved, or perfected, extra-neural means by which, without altering our nervous system, we can refine its operation and expand its scope.[4]

A civilization makes progress by leveraging the achievements and observations of past generations. We compress history into words, stories, and symbols that allow living people to learn and benefit from the experiences of the dead. In the space of one childhood, we can learn what it took humanity many centuries to figure out. While animals may have some capacity to instruct their young, humans are unlimited in their capacity to learn from one another. Thanks to stories, books, and our symbol systems, we can learn from people we have never met. We create symbols, or what Korzybski calls abstractions, in order to represent things to one another and our descendants more efficiently. They can be icons, brands, religious symbols, familiar tropes, or anything that compresses information bigger than itself.

And unlike animals, who can't really abstract at all, the number of abstractions we humans can make is essentially limitless. We can speak words, come up with letters to spell them and numbers to represent them digitally. We can barter objects of value with one another. We can trade for gold, which represents value. We can trade using gold certificates, which represents the value of gold. We can even trade with modern currency, which represents value itself. Then, of course, we can buy futures on the value of currency, derivatives on the value of those futures, or still other derivatives on the volatility of those.

Thanks to our abstract symbol systems, we can see a 28° in the corner of the TV screen and know to wear a coat. Everything from the ability to record temperature to the technology to broadcast images has been leveraged in order for us to make the split-second decision of what to wear. The bounty of thousands of years of civilization is reaped instantaneously. That's the beauty of the short forever.

Of course, all this abstraction is also potentially distancing. We don't see the labor that went into building our railroads or the civilizations that were wiped out in order to clear the land. We don't see the millennia of dinosaurs or plankton that went into our oil, the Chinese repetitive stress injuries that went into our iPhones, or any of the other time-intensive processes we can spend in an instant today. We tend to see math and science as a steady state of facts rather than as the accumulated knowledge of linear traditions. As Korzybski put it, we see further because we "stand on the shoulders"[5] of the previous generation.

The danger of such a position is that we can forget to put our own feet on the ground. We end up relating to our maps as if they were the territory itself instead of just representations. Depending solely on the time binding of fellow symbol makers, we lose access to the space binding of our fellow animals, or the energy binding of nature. Like overscheduled workers attempting to defeat the circadian rhythms through which our bodies self-regulate, we attempt to operate solely through our symbol systems, never getting any real feedback from the world. It's like flying a plane without a visible horizon, depending solely on the information coming from the flight instruments.

What we really need is access to both: we want to take advantage of all the time that has been bound for us as well as stay attuned to the real-world feedback we get from living in the now. While they often seem to be at odds, they are entirely compatible, even complementary, if we understand the benefits and drawbacks of each. It's not solely a matter of establishing appropriate "temporal diversity," as Stewart Brand suggests. That may work for processes that are unfolding over historical time, but it doesn't really have anything to do with the *now* that we're contending with in a presentist reality. We're not in nature's time nor fashion's time. We're in *no* time. All we presentists get from zooming out to ten-thousand-year time spans is vertigo.

The stuff of time binding—all that information, however dense—is like the data on a hard drive. A presentist life takes place in something more like RAM—the active memory where processes are actually carried out.* The shift from a historical sensibility to a presentist one is like being asked to shift our awareness away from the hard drive and into RAM. It's all processing, with no *stuff* to hang on to.

A lot of what ends up feeling like overcompression and data deluge comes from expecting RAM to act like a hard drive, and vice versa. It's like depending on the Vatican for progressive values, or on a high-tech start-up for institutional ballast. We simply can't use one to do the other without unbinding stored time or binding up the present. Instead of a contemplative long now, we get an obsessive short forever.

///

SEEDS AND FEEDS

The concept of the long now too easily fails us because it compares different *scales* of time instead of different *kinds* of time. All the

---

* Yes, when a computer runs out of active memory, special compensatory processes can take over that use extra space on the hard drive as makeshift RAM—virtual RAM. But the grinding sounds a hard drive makes under such stress, and the tendency for programs to run slow and hang under such conditions, just underscores how much better it is to use the appropriate kind of hardware for the job. Interesting, especially in a presentist, always-on world, RAM has proved much better at working as a hard drive than the other way around.

layers of pacing explored by Dyson and Brand are still aspects of the same linear process. They are just proceeding at different rates. To be sure, attending to affairs of government at the rate of fashion or addressing the needs of the planet at the rate of commerce produces problems of its own. Presidencies end up driven by daily polling results, and the future costs of pollution never find their way onto the corporate balance sheet. But these are all aspects of a cultural acceleration that has been in progress for centuries.

Now that we have arrived in the present, we can finally see that the real schism today is less between these conflicting rates of time than it is our conflation of two very different relationships to time. There is stored time—the stuff that gets bound up by information and symbols. And then there is flowing time—the stuff that happens in the moment and then is gone. One needs to be unpacked. For the other, you have to be there.

Stored time is more like a pond than a stream. It remains still long enough to promote life and grow cultures within it. A pond may be stagnant and unsuitable for drinking, but that only attests to its ability to support a living ecosystem within itself. A stream, on the other hand, is defined by its constant movement. It is never still. This doesn't mean it lacks power. Over time, its flow can cut a path through solid rock. But it's a hard place for cultures to develop. The pond creates change within itself by staying still. The stream creates change beyond itself by remaining in motion. If we think of them as media, the pond contains its content, while the stream uses the earth around itself as its content.

Likewise, our informational content comes to us both as ponds and streams—stored data and flows of data. The encyclopedia is relatively static and stored, while the twenty-four-hour news channel is closer to flow. Yes, the encyclopedia may change every few years when a new edition comes out, and the news channel may broadcast some canned video reports. But the value of the encyclopedia rests in the durability of its assertions and the cumulative authority of its institutional history. The value of the twenty-four-hour news channel is based in the freshness of its data—the newness of its news.

Where we get into trouble is when we treat data flows and data storage interchangeably. This is particularly easy to do in digital environments, where even fundamentally different kinds of information and activities are rendered in ways that make them all look pretty much the same. So we scan a digital book or article with the same fleeting attention as we regard a Twitter stream or list of Facebook updates.

This means we are likely to race through a longer text, hoping to get the gist of it, when the information is just too layered to be appreciated this way. By rushing, we relegate deep thoughts to the most transient, temporary portions of our memory and lose the ability to contemplate anything. We are overwinding, in that we are attempting to compress a lengthy, linear process into a single flow moment.

Or just as futilely, we attempt to catch up with a Twitter feed, as if yesterday's Tweets needed to be absorbed and comprehended like the missed episode of a television series. Instead, Twitter needs to be embraced as the stream it is—an almost live flow of facts and commentary whose relevance is conditional on the moment. Twitter is how we complain that there was no instant replay of a questionable pass reception, how we share our horror about a school shooting that just occurred, how we voice our solidarity with a protest in progress, or how we let other protesters know where the cops are stationed. Catching up with Twitter is like staying up all night to catch up on live streaming stock quotes from yesterday. The value was in the *now*—which at this point is really just a *then*.

Stored information, like a book, is usually something you want to absorb from beginning to end. It has greater longevity and is less dependent on the exact moment it comes out. We can read it in our own time, stopping and starting as we will, until we get to the end.

Flowing information, like twenty-four-hour news or MTV videos, is more like the nonnarrative experience of electronic music or extreme sports. We get a textural experience, we learn the weather, or we catch the drift. We do not get to the end; we shut it off and it continues without us.

Where it gets even harder is with information types that have the qualities of both storage and stream. An email inbox may look like a Twitter stream, but it's not quite the same beast. Yes, they are both lists that scroll down the screen in chronological order. However, individual emails are more like stored information, while the Twitter stream is pure flow. Even if you receive a new email every minute, opening a message takes you out of the current and into a static tide pool. For the most part, the content within the message won't continue to change. It will stay still long enough for you to take it in, form impressions about it, and even begin to compose a response. In theory (although for many of us, not in practice anymore), email is something that can be caught up with. We return to our inbox after however many minutes, hours, or days away, and then deal with whatever is inside. Of course, in that single session of email we probably won't actually complete the tasks demanded by each message. Or even open every one. But we will at least scan the headers, deal with whatever is urgent, and make mental notes of the rest.

It's easy to reverse email's biases, however, and to treat the list of messages like stored data and the messages themselves like flow. Some workflow efficiency experts have suggested that people strive for something called "inbox zero"[6]—the state of having answered all of one's emails. Their argument, based on both office productivity and cognitive science, is that merely checking one's email is inefficient. If the email is not processed—meaning answered, deleted, filed, or acted upon in some way—then it remains part of a growing to-do list. According to this logic, the time spent checking it was wasted if the message hasn't been processed to the next stage.

Worse, a new loop has been opened in the brain—there's one more thing weighing on the consciousness.

Backing all this up is the idea that mental stress is the result of opening up new unresolved tasks—like problem tickets on a service department's desk. Every individual thing we haven't figured out sits on our awareness like a live, ticking clock. Every unanswered question and every task we haven't yet scheduled stays in the most active part of our brain, waiting for an answer. The way to reduce this mental stress is to close as many of these open, running loops as possible. This doesn't necessarily mean accomplishing every task that we have taken on, but rather being able to visualize when and how we are going to do it. If your spouse asks you to buy some milk, this task stays open until you make a mental note to pick it up on the way home. You make sure you have enough money, you remind yourself that the store will still be open at that hour, and you consider what route will take you the least out of your way. Once you know how you're going to accomplish the task, the loop is closed— even though the task has not yet been accomplished. There's no more that can or need be done in the present, so the active part of the brain is freed up. This is a proved method of reducing stress.

The inbox zero people see each message as a running loop. So they recommend we do something to create closure for each email— answer it, put a date on our calendar, add something to our to-do list, or even just delete it—rather than just leaving it sit there. According to *Bit Literacy* author Mark Hurst, if we don't get our inbox empty, we won't get that "clean feeling."[7] Given that the email inbox will nearly always refill faster than we can empty it, and that the messages arrive on everyone else's schedule rather than our own, this clean feeling may be short-lived or even unattainable. Instead, this quest for cleanliness becomes an obsessive-compulsive loop all its own.

Looked at in terms of flowing and static information, the email inbox is one, big, unfinishable loop. It is not a book or document that can be successfully completed. It is a flow. Sure, we can mark or move emails that are important, create priorities and sorting routines. But the initial choice to have email at all is to open a loop. The choice to open a particular email, though, constitutes entry into something more like static information. The problem is that the sender may have spring-loaded a whole lot of time and energy into that message, so that clicking on it is like opening a Pandora's box of data and responsibilities. A week of the sender's preparation can instantaneously unfold into our present.

Think of how much less of an assault it may feel like if the sender simply enclosed all that spring-loaded time in a document and then attached it to that same email for you to file for later. Or even better, a link to a document online somewhere else. Of course, the savvy computer user can file away the offending email almost as easily as an attached document. But he is simply relegating information to storage that was misrepresented as flow.

Such distinctions may seem trivial in the realm of email, but the inability to distinguish between flow and storage in other contexts can make or break businesses and economies. For just one example, take the medium of money. Or, since there were actually many kinds, the media of moneys.

Back in the late Middle Ages, the rigid power structures of feudalism were shaken by the emergence of locally minted currencies based on grain. For close to a thousand years before that, peasants knew nothing but subsistence living, in absolute obeisance to the lord whose fields they tended. Time had been standing still; the flow of history was essentially halted as aristocratic families maintained their fiefdoms. That's why this era was often called a dark age.

There was a bit of barter between people, but this was a slow and inefficient form of trade. A person who had chickens but wanted shoes would have to find a shoemaker who wanted chickens. If only people had one thing of agreed-upon value they could use to trade, then everyone could get the things they wanted. Grain receipts provided people with just that form of early currency. Farmers brought their seasonal harvests to grain-storage facilities, where they were given receipts for how much grain they had deposited. A hundred pounds of grain would be acknowledged with a stamped, paper-thin foil receipt, perforated into smaller sections. Holders of the receipts could tear off little sections and use them to buy anything else at market. Even people who did not need grain could use the receipts, because the value was understood, and eventually someone would actually need grain and be able to claim the specified amount.

Now, while we may think of these grain receipts as a storage of value, they were, in fact, biased toward flow. Their purpose was less to store the value of the grain than to monetize and move it; they allowed for people to transact on something that would otherwise be stuck in dead storage. Pushing this de facto local currency even further to the flow side was the fact that these receipts lost value over time. The grain storer had to be paid for his services, and some of the grain would inevitably be lost over time to rodents and spoilage. In order to compensate for this, the value of a grain receipt would be lowered at regular intervals. One year it would be worth ten pounds; the next it may be worth nine. So it was to everyone's advantage to spend the money rather than to hold it.

And spend they did. Money circulated faster and spread wider through its communities of use than at any other time in economic history.[8] Workers labored fewer days and at higher wages than before or since; people ate four meals a day; women were taller in Europe than at any time until the 1970s; and the highest percentage on record of business profits went to preventative maintenance on equipment. It was a period of tremendous growth and wealth. Meanwhile, with no way of storing or growing value with this form of money over the long term, people made massive investments in architecture, particularly cathedrals, which they knew would attract pilgrims and tourists for years to come. This was their way of investing in the future, and the pre-Renaissance era of affluence became known as the Age of Cathedrals.

The beauty of a flow-based economy is that it favors those who actively create value. The problem is that it disfavors those who are used to reaping passive rewards. Aristocratic landowning families had stayed rich for centuries simply by being rich in the first place. Peasants all worked the land in return for enough of their own harvest on which to subsist. Feudal lords did not participate in the peer-to-peer economy facilitated by local currencies, and by 1100 or so, most of the aristocracy's wealth and power was receding. They were threatened by the rise of the merchant middle class and the growing bourgeois population, and had little way of participating in all the sideways trade.

The wealthy needed a way to make money simply by having money. So, one by one, each of the early monarchies of Europe outlawed the kingdom's local currencies and replaced them with a single central currency. Instead of growing their money in the fields, people would have to borrow money from the king's treasury—at interest. If they wanted a medium through which to transact at the local marketplace, it meant becoming indebted to the aristocracy.

Unlike local, grain-based currencies, these central currencies were not biased toward flow, but toward storage. In the new system, having money meant having a monopoly on transaction. Those who wanted to transact now needed to borrow money from the treasury in order to initiate their business cycles. Only those with large amounts of scarce capital could lend it, and they did so for a premium. Hoarding money was no longer a liability but the surest means to greater wealth.

The shift to central currency not only slowed down the ascent of the middle class, it also led to high rates of poverty, the inability to maintain local businesses, urban squalor, and even the plague. Over the long run, however, it also enabled capitalism to flourish, created the banking industry, and allowed for European nations to colonize much of the world.

Our current economic crises stem, at least in part, from our inability to recognize the storage bias of the money we use. Since it is the only kind of money we know of, we use it for everything. We naturally tend to assume it is equally good at flow and storage, or transaction and savings—but it's not. That's why injections of capital by the Federal Reserve don't end up as widely distributed as policy makers imagine they will be. It's also why government can't solve local economic depression simply by getting a bank to lend money for a corporation to open a plant or megastore in the afflicted area. Failing local economies need flow, not more storage. Policies encouraging local peer-to-peer transactions end up allowing people to create value for one another, and a local economy to sustain itself the old-fashioned way. Top-down, central currency simply isn't the very best tool for that job. In Japan, for example, when the greater economic recession seemed intractable, current and former government officials encouraged rather than discouraged local and alternative currency innovation. This resulted in over six hundred successful currency systems, most famously a trading network called Fureai Kippu, through which hundreds of thousands of people earn credits to pay for care of elder relatives.[9]

Of course, the net has opened the way for more flow. Capitalism's crisis today might even be blamed on the net's ability to birth new businesses with almost no investment. Two kids with a laptop and virtually no capital can create and distribute music, television programming, or smart phone apps capable of earning millions of dollars. What's a venture capitalist to do? On an innovation landscape now characterized by flow, capital competes for vehicles in which to invest. Accordingly, interest rates plummet and banks threaten insolvency.

Instead of borrowing money from banks, depressed communities can jump-start their local economies by using alternative currencies biased toward flow. Time Dollars and local economic transfer systems (LETS), to name just two, allow people to list their skills or needs on a website, find one another, and then pay in locally defined units, or even "hours," of time. Everybody starts with the same amount, or a zero balance, and goes up or down as they receive or provide goods and services. There's no incentive to hoard currency, since it is good only for transacting. People just try to maintain equilibrium. It is present-based currency, encouraging transactions in the here and now.

While alternative currencies have yet to solve all our economic woes, they reveal the inadequacy of a storage medium to solve problems of flow and vice versa. When we attempt to pack the requirements of storage into media of flow, or to reap the benefits of flow from media that locks things into storage, we end up in present shock.

MASHUP AND MAKEUP

The women on *The Real Housewives of Beverly Hills* hurt one another's feelings quite a lot. Every episode seems to hinge on a misunderstanding between two or more housewives over a meal, which is then amplified via text message and Facebook, and ultimately ends in a full-fledged fight. Unlike those on *Mob Wives* or one of the New Jersey reality series, these fights don't take the form of physical, hair-pulling brawls, but there is nonetheless something intensely and unnervingly physical about the nature of these conflicts.

The only other thing characterizing the series—besides the toxic wealth of its subjects—is plastic surgery. All the people on this program wear faces frozen in time by their many procedures. The skin around their eyes is stretched tight, hiding expression lines. The women's lips are infused with weighty collagen, muting what may be smiles and leaving their mouths in the half-opened stupor of a person with a chronic sinus infection. And their foreheads are quite literally paralyzed by botulism injections, leaving eyebrows in the locked and angular orientations of their last facelifts.

They seem to stare in wide-eyed disbelief at everything everyone says, but that's only because they are staring, wide eyed, all the time, by default. They have no choice; they cannot blink. In the quest to lock in their thirtysomething looks, they also locked their faces in a permanent, confused glare.

No wonder they have so many misunderstandings; they are missing out on that 94 percent of human communication that occurs nonverbally. It's not only the words we say but the visual cues we send while saying them. The tension in our mouth, the shape of our eyes, the lines in our foreheads, and the direction of our eyebrows tell what we are feeling. How else to distinguish between enthusiasm and sarcasm, or a joke and a complaint? Sometimes our faces communicate indirectly, compensating unconsciously for our words. Threatening statements may be undermined by a softening gaze or subtle smile. It's as if our bodies know how to keep the peace, even when our words may be disagreeing.

The Housewives have no access to this dashboard of human expression. They have traded their ability to communicate and commune in the present moment for an altogether unconvincing illusion of timeless beauty. In attempting to stop the passage of time and extend the duration of youth, they have succeeded only in distancing themselves from the moment in which their real lives are actually transpiring. Incapable of engaging and connecting with other human beings in real time, they send each other false signals. Their faces don't correspond to the ideas and emotions they are expressing, and they appear to be lying or covering up something. Or their silent signals seem inappropriate or even numb to the thoughts and expressions coming from others.

They are trapped in another version of the short forever—one in which a particular stage of one's life is deemed to be better than the rest, and so everything before and after is remade in its image. Twelve-year-olds and forty-year-olds both aim for about age nineteen, making children look promiscuous and women look, well, ridiculous. And as any nineteen-year-old knows, it's not actually a particularly perfect moment in life to celebrate above any other. In fact, as fashion magazines, advertisers, and fictional media focus on this age as the pinnacle of human experience, late adolescents feel even more pressure to achieve physical, sexual, and social perfection. This is not a reasonable expectation for a college sophomore; 25 percent of them engage in bingeing and purging as a weight-management technique,[10] and nineteen is the median age for bulimia,[11] which may as well be the signature disease for overwinding and subsequent explosion.

We have all stuck our heels in the ground at certain moments in our lives, attempting to slow the progress of time, only to get whipped back around to our real ages once we lose our grip. These temporal compressions generate an almost spasmodic movement through time, where the trappings of one moment overaccumulate and prevent our moving on to the next. It's present shock as Peter Pan syndrome, where the values of youth are maintained well into what used to pass for adulthood.

A growing number of names are emerging to identify these temporally compressed lifestyle choices. "Grup," for example, is *New York* magazine journalist Adam Sternbergh's new term for the aging yuppie hipster. Grups is the word for "grown-ups," used on a world ruled by children in a *Star Trek* episode: turns out all the adults have died of a strange virus that quickly ages and kills anyone who has passed puberty—but extends the lifetimes of children to hundreds of years. Likewise, real-world grups are hip, indie fortysomethings who, according to Sternbergh, "look, talk, act, and dress like people who are 22 years old. It's not about a fad but about a phenomenon that looks to be permanent."[12] Grups wear the vintage sneakers of their own childhoods, put their babies in indie rock T-shirts, and use messenger bags instead of briefcases.

Eventually, the time compression takes its toll, requiring some pretty intense mental gymnastics: "If you're 35 and wearing the same Converse All-Stars to work that you wore to junior high, are you an old guy sadly aping the Strokes? Or are the young guys simply copying you? Wait, how old are the Strokes, anyway?"[13] Just as in the *Star Trek* episode, the realization of one's adulthood comes on suddenly, painfully, and fatally, as the clock's overwound mainspring suddenly releases, long overdue.

Hipsters—the fashionably antifashion young adults found in artsy neighborhoods, from Brooklyn's Williamsburg to San Francisco's Mission or Lower Haight—suffer from the reverse distortion. They appropriate the styles of previous generations and pack them into the present in order to generate a sense of timeless authenticity. Unlike grups, they are actually young, but operating under the tacit assumption that creativity and authenticity lie somewhere in the past. They may drink Pabst beer and wear Dickies—both brands with rich histories, such as brand histories go. But the hipsters are unaware that the many companies they feel they have discovered are utterly aware of their own repositioning and revival as hip retro brands. Zippo lighters and V-neck T-shirts may reference a moment in American working-class history, but purchasing them as fashion items is superficial at best—not a real affiliation with the proletariat. That they provide a sense of grounding and reality to young people today says less about the high quality and authenticity of mass-produced mid-century goods than it does about the untethered, timeless quality of the hipster experience. Authenticity comes to mean little more than that an object or experience can be traced to some real moment in time—even if it's actually being purchased at Walmart or on Amazon.com.

The ready availability of anything to anyone removes it from whatever its original context may have been. In the mid-twentieth century, cultural theorist Walter Benjamin wondered about the effects of mechanical reproduction on the work of art. What would it mean that people could see reproductions of paintings in books or listen to music on records without ever beholding the work for real, in their original settings? Would the aura of the originals be lost?

As anyone who grew up in the twentieth century knows, however, to discover and learn about a counterculture or art movement still involved a journey. Only certain used-record stores and bookshops may carry the genre in question, and even knowing what to look for required mentorship from those who went before. And getting that help meant proving oneself worthy of time and tutelage. Beat poetry, psychedelic literature, Japanese pop music, and John Cage recordings weren't available via a single Google search. Accumulating knowledge and content took time, and that time was a good and necessary part of the experience. It didn't simply make these nooks and crannies of culture more elitist; it helped keep them more like tide pools than oceans. Their stillness and relative obscurity helped these genres grow into unique cultures.

When everything is rendered instantly accessible via Google and iTunes, the entirety of culture becomes a single layer deep. The journey disappears, and all knowledge is brought into the present tense. In the short forever, there is no time to prepare and anticipate. No wonder people hang on to the musical styles and fashions of their youth. Finding them took a kind of time—a particular windup— that is unavailable to cultural explorers today.

It is also unavailable to the cultural creators. No sooner is a new culture born than it is discovered by trend-setting *Vice* magazine; covered by the *New York Times* Style section; broadcast on MTV; and given a book, record, or movie deal. There is no time for an artist or scene to develop unless those involved take extreme measures to isolate themselves and avoid being noticed. As a result, there is no time to develop the layers and experiences required for a genre to evolve.

As a substitute for this process, temporal compression takes the form of mashup. Originally just a way of describing how a deejay may overlay the vocal track of one recording onto the instrumental track of another, mashup now refers to any composition or process that mashes up previously distinct works into something new. Mashup is to culture as genetic engineering is to biological evolution. Instead of waiting to see how genres merge and interact as cultures over time, the artist cuts and pastes the cultural strains together.

A lot of it is tremendous fun—such as a YouTube video that mashes up Nirvana's "Smells Like Teen Spirit" with Boston's "More Than a Feeling,"[14] demonstrating how the two songs are nearly identical. But a lot of mashup is also serious, thought-out art. *The Grey Album*, released by Danger Mouse in 2004, mashes up an a cappella version of rapper Jay-Z's *Black Album* with instrumentals from the Beatles' *White Album.* It not only led to lawsuits over its use of copyrighted Beatles' content, but also won the critical attention of the *New Yorker* and Album of the Year from *Entertainment Weekly.* Mashup mix-artist Gregg Gillis's project Girl Talk has released records that are both commercial and critical hits. The *New York Times* waxes almost reverential: "Girl Talk has created a new kind of hook that encompasses 50 years of the revolving trends of pop music. Sometimes cynicism is a hook, sometimes the hook is humor, angst, irony, aggression, sex or sincerity. Girl Talk's music asserts all these things at once."[15] The Museum of Modern Art in New York regularly features exhibitions and performances by mashup artists, such as DJ Spooky remixing D. W. Griffith's footage with modern imagery into a new piece called *Rebirth of a Nation*. While mashups may compress time, they do allow for a new sort of commentary, intention, and irony to emerge.

In the twentieth century, cubism responded to the deconstructed processes of the Industrial Age by expressing everything as deconstructed shapes. Artists broke from the tradition of showing something from a particular perspective and instead used multiple points of view in the very same painting. The artist pulled apart objects into planes that would normally be visible only from multiple vantage points. This allowed them to show more than one facet of a person or scene at the same time. While cubism looked at one moment from different perspectives, mashup looks at one perspective from multiple moments. Perhaps best understood as cubism's twenty-first-century corollary, mashup accomplishes the reverse: instead of sharing one moment from multiple perspectives, it brings multiple moments into a single whole. Twenties jazz, sixties rock, nineties electronica—all occurring simultaneously. Where cubism compresses space, mashup compresses time. Cubism allowed us to be in more than one place at the same time; mashup allows us to be in more than one time at the same place.

Although it may not be as classically structured or emotionally resolved as the artistic works of previous eras, mashup does express the temporal compression almost all of us feel on a daily basis. Fifteen minutes spent on Facebook, for example, mashes together our friendships from elementary school with new requests for future relationships. Everything we have lived, and everyone we have met, is compressed into a virtual now. While grade school relationships used to be left back in childhood, they reemerge for us now—intentionally forgotten memories forcibly shoved back into current awareness. We live all of our ages at once. Nothing can be safely left behind.

For digital memory never forgets. After conducting a simple Google search, US immigration agents at the Canadian border denied entry to Andrew Feldmar, a seventy-year-old college professor, because they found an obscure reference to the fact that he had taken LSD in the 1960s. While that case was sensational enough to make the headlines, this same inability to shed the past from the present affects us all. According to research by Microsoft, 75 percent of human resources departments do online research about their candidates, utilizing search engines, social networking sites, personal blogs, and even photo-sharing sites. Seventy percent of these prospective employers say they have rejected candidates on the basis of their pictures and comments.[16] And our own online hygiene may not be enough to spare us the consequences. Did someone else snap a photo of you while you were drunk at her party? Did she upload the picture to Facebook (or did her camera do it automatically, as so many smart phones now do)? Facial-recognition software online can tag the photo with your name if someone doesn't do that manually. And even if you find out and get it deleted, it is always there in someone's history or hard drive. Moreover, Facebook pages can be temporarily hibernated, but they cannot be removed.

This ends up favoring the past over the present. Most societies give individuals the opportunity to reinvent themselves over time. We are either forgiven for past mistakes, or they are eventually forgotten. Jewish Talmudic law requires people forgive all trespasses at regular intervals, and even forbids someone from reminding another of an embarrassing moment from his past or childhood. The ancients understood that community could not function if knowing someone for a long time was a social liability instead of a strength. In more recent history, written records would often be expunged after a period of time, a person could move to a new neighborhood and start over, or a bankruptcy cleared after seven years.

Today the new permanence of our most casual interactions—and their inextricability from more formal legal, financial, and professional data about us—turns every transient thought or act into an indelible public recording. Our résumés are no longer distinct from our dating histories. It's not just the line between public and private activity that has vanished, but the distance between now and then. The past is wound up into the present and no longer at an appropriate or even predictable scale. The importance of any given moment is dependent solely on who has found it and what they use it for. Many of the top political science and constitutional law graduates steer clear of politics or even judgeships for fear of the scrutiny that they and their families will be subjected to. Nothing, no matter how temporally remote, is off-limits. A forgotten incident can resurface into the present like an explosion, threatening one's reputation, job, or marriage.

Our recorded past then competes with our experienced present for dominance over the moment. In his book *Delete*, Viktor Mayer-Schönberger tells the story of a woman who met up with an old boyfriend and made a date, thinking she might rekindle the relationship—particularly now that so much time had passed since their breakup. But she went ahead and reviewed her old email exchanges with him, archived deep in her trash folder. It brought the painful memories back into the present, and she canceled the date. However she or her former beau may have changed or grown since then—whatever connection had been forged between them when they actually reunited in the real world—was rendered irrelevant in comparison with the past magnified through technology.

Even our Facebook identities can now be unwound as timelines depicting our previous states on the site—the friends we later unfriended, the silly movies and books in which we were once interested, and the embarrassing things we said and did that we wish would recede into the distance. But in the short forever, nothing recedes. Everything relative is now also relevant.

What isn't coming at us from the past is crashing in at us from the future. Our Facebook profiles and Google accounts become parts of big data sets, where they can be mined for patterns and modeled against those of millions of other users. The computers at companies like the market research firm Acxiom and big data analytics specialists Opera Solutions churn data not simply to learn what we have done, but what we *will* do. Opera is not merely doing analytics, but what they call predictive analytics. As they put it: "We have created Vektor™—a secure and flexible Big Data analytics platform that extracts powerful signals and insights from massive amounts of data flow, and then streams analytically enriched guidance and recommendations directly to the front lines of business operations."[17]

It doesn't require smarts, just computing brawn. Big data companies collect seemingly innocuous data on everyone, such as the frequency of our text messages, the books we've bought, the number of rings it takes us to pick up the phone, the number of doors on our cars, the terms we use in our Web searches, in order to create a giant profile. They then compare this profile against those of everyone else. For reasons no one understands, the data may show that people who have two-door cars, answer the phone in three or more rings, and own cats are extremely likely to respond favorably to ads for soup. So these people will be shown lots of soup advertisements. The market researchers don't care about the data points themselves or the logic connecting one behavior to another. They care only about predicting what a person is statistically likely to do. Privacy has nothing to do with a particular fact they may know about you, but things they know about you that you may not even know yourself.

Now that most of us have migrated to the Internet, the number of data points available to big data companies is almost infinite. Every mouse click and navigational pathway, the lengths of our incoming and outgoing emails, the amount of time we spend with more than one window open, and so on, may not give any logical clues about our inner workings, but how they stack up against the same data points of others means everything. Comparing and contrasting all this data, modelers can identify the future sexual orientation of prepubescents, the probability of our requiring fertility treatments, the likelihood of our changing political party affiliations, or even if we're about to come down with the flu. And with startling accuracy, they are correct.

So it's not only our past, but our futures that are compressed into the present, as well. We end up in a short forever—a psychic mashup—filled with contradiction and paralyzed by both the weight of an indelible history and the anticipation of a preordained fate.

ACT NOW

Black Friday gets worse every year. The infamous day-after-Thanksgiving sale-a-thon, when stores launch the Christmas shopping season by offering their deepest discounts of the year, seems to get more extreme, more urgent, and more violent each time it comes around. The stakes are high, and not just for the consumers who trample one another—sometimes to death—in order to get through Walmart's doors the second they open. Equally invested are many market analysts and economists, who now treat Black Friday's results as reliable indicators of the nation's financial health.

Fully aware that the Black Friday sales figures will lead the headlines and have a significant impact on the following Monday's stock indexes, investors look to what's happening the day *before* Black Friday for a hint of how to be positioned for the actual Black Friday, which is really just a way to be positioned for whatever happens on Christmas. Knowing they are being judged in advance and eager to get a jump on their competitors, retailers edge toward increasingly earlier opening times. While the earliest Black Friday sales used to begin Friday morning at 9 a.m., by the early 2000s they had moved up to 6 a.m. or even 5 a.m. Customers lined up in the cold outside their favorite big-box stores on Thursday night, and local news shows showed up to cover the spectacle.

By 2011 some of the most aggressive stores, such as Target, Best Buy, and Macy's, decided that they would push the envelope even further and start Black Friday at midnight. Walmart rolled Black Friday all the way back to Thursday evening at 10 p.m. Shoppers showed up, but now they were complaining. Some were upset that they were being required to leave their families during Thanksgiving dinner in order to get a good place on line. Others felt the expanded hours just lengthened the shopping day beyond their endurance levels. Some even seemed aware of their complicity in overworking store clerks, and of how the fun of Black Friday had turned into more work for everyone.

Employees complained, too, and those at some of the big-box stores were fired for refusing to come in for the overnight Thanksgiving shift. Memories of late-nineteenth-century union fights over workers' hours were retrieved by the press: "Even though it's a desperate time doesn't mean that we should trade all that ground that our fathers and our grandfathers, everyone that came before us, fought really hard for," a Target worker told the *New York Times*.[18] JCPenney, in a nod to these sentiments, kept their opening time at a respectable 4 a.m., because "we wanted to give our associates Thanksgiving Day to spend with their families."[19]

The extreme overwind has pushed many shoppers and workers over the edge, and even threatened the Christmas shopping season as a whole. What was once a seasonal consumer sport now feels to many like work and an intrusion on the rest of the holiday. Having found a temporal anchor in the limbo of generic, big-box shopping, retailers couldn't help but spring-load it until it just broke. Starting the Christmas season the day after Thanksgiving was already at the very boundary of spring-loading; pushing into Thanksgiving itself was an overwind. It broke through the patina of holiday spirit, masking this otherwise crude effort to get people to go further into credit card debt by encouraging them to purchase more electronics and other goods manufactured in Chinese plants and sold in big-box stores that kill local business. All this, we must remember, on borrowed money and borrowed time. No wonder our consumer economy went into present shock.

In the process, many consumers and workers alike came to realize the artificiality of the whole affair and simply turned away. Anticonsumerist *Adbusters* magazine's "Buy Nothing Day" had already morphed into the lifelong commitment to Occupy Wall Street. This new wave of local, presentist values only reinforced the breakdown between people and the consumption they were obligated to be doing to save America's corporations from peril.

Black Friday is just a more condensed version of the time shifting that characterizes our entire consumer economy and its current crisis. It is incumbent upon businesses to unwind the temporal compression inherent in consumer culture in order to transcend it. In short, we must come to understand where various forms of pressurized time are doing more harm than good.

On its most fundamental level, a consumer society is based on its ability to compress the time of others through mass production. Issues of exploitation and ethics aside, mass production and Industrial Age technologies were, at their core, about compressing the labor of faraway people into the products used at home. While, as we saw, peasants were originally able to lift themselves beyond subsistence farming by learning crafts and trading goods, the Industrial Age took this a step further. Along with their invention of central currency, the waning aristocracy of the late Middle Ages also created the chartered monopoly. The charter was a legal contract between the monarchy and a company, granting it the exclusive right to do business in a particular region or industry. A company that was granted a charter no longer had to worry about competition or going out of business due to an unforeseen setback such as bad weather or a shipwreck. The monopoly enjoyed permanence, by law. In return for this favor, the company would give the king or queen some shares.

So these big companies, with names like British East India Company or the Dutch East India Trading Company, went to the faraway places to which they owned all rights and brought back resources that could then be fabricated into clothes, foods, and other products in local factories. It was a neat trick—which robbed most everyone of the ability to do business in the present. People who once purchased locally produced goods now had to buy them from the chartered monopolies. Moreover, people who once created goods now had to go work for a company instead. The present-based reality of a normal local economy ended up superseded by something else.

This was also a perfect environment for the Industrial Age to take hold. Machines didn't simply make things faster; they allowed companies to hire less expert labor. Where a skilled shoemaker was once required to cobble a shoe, now shoes could be assembled by machine operators who each needed to know only how to perform one small step in a shoe's manufacture. Such laborers could be found easily, paid much less, and fired without cost, for they didn't take any expertise away with them.

The expertise a real cobbler, or any skilled craftsperson, may have accumulated over years of apprenticeships was rendered worthless, as the time formerly invested in human training was now compressed into the mechanical processes of the machine itself. Yes, a skilled shoemaker might be required to help devise the machine in the first place, but after that he is expendable—particularly for a company that has no need to innovate, because its stature is guaranteed by its monopoly. So on the simplest level, the time compression that once allowed a craftsperson to leverage his experience is now transferred to the machine. The worker is always in the present, no more valuable on his hundredth day than on his first.

When looked at in terms of value creation, the whole industrial model starts to break down. American colonists, for example, created value by growing cotton in their fields. They were unable to reap the market value of their product, because they were forbidden by law from selling it to other colonists. They were required instead to sell it at fixed prices to the British East India Company, who shipped it back to England, where it was fabricated into clothes by another monopoly, and then shipped back to the colonists to purchase at a premium. This was not more efficient for anyone; it simply prevented anyone but a chartered monopoly from adding value to (and making money from) the cotton. America's colonists were cut out of the value-creation equation and subjected instead to a system where their hard labor in the present was simply loaded into the production cycle of a long-term and long-distance corporate product.

America's white colonists, or at least the educated people who led them, knew better than to accept these terms and fought a war of independence. Britain learned its lesson—but maybe not the one we would have hoped for. What they realized was that their methods of value extraction worked a whole lot better on enslaved indigenous populations who had neither the experience to envision nor the gunpowder to choose otherwise.

By around the time of Queen Victoria, colonial empires had established corporations overseas that both mined resources and fabricated them into products. These were completely closed circuits. Britain sent mechanical weaving looms to India, for example, and maintained their monopoly over production simply by forbidding hand weaving. The only way to weave was to work for one of Britain's loom owners. While opinions vary today on exactly how this impacted India's capacity to make money over the long term, we tend to think less about the way this impacted the habits of every-one back at home.

While people may not have been particularly concerned about Britain's exploitative trading practices, they were not accustomed to purchasing outsourced goods. In 1851 Queen Victoria and the cor-porations she sponsored held the Great Exhibition of the Works of Industry of All Nations, in a high-tech, million-square-foot glass pavilion called the Crystal Palace. It was like a world's fair for the Industrial Revolution, where fourteen thousand exhibitors demon-strated their factory-made wares and the machines that created them. Spectators marveled at the steam hammers, hydraulic presses, ba-rometers, and diving suits of this new, mechanized era.

As I argued in my book *Life Inc.*, the Great Exhibition's primary intent was to distract the domestic public from the dark underbelly of international industrial modernity. Through this spectacle, Queen Victoria and the corporations she sponsored disconnected these tech-nologies from the human toll they inflicted on their operators. As if in a shopping mall, people gawked, their jaws dropping, at the steam pipes and gears, utterly unaware of the faces and hands the machines burned and mauled. People saw products and production, but never the producers themselves. If anything, industrial modernity was simpler and cleaner than manual labor. As much a welcome step back as a daring leap forward.[20]

The Great Exhibition was designed to convey precisely this sen-sibility and to help the British contend with the shock of the new. Organizers cleverly promoted and organized the event as a celebra-tion of faux-medieval design and dedicated the central hall to an ex-hibit on Gothic revival architecture. This was part of a larger effort to disguise the industrialization of Victorian England as a throw-back to feudal tradition. The era of the high-tech factory would be rebranded as the romantic revival of medieval monarchy. The Great Exhibition mythologized both free trade and the Industrial Age as a return to the best of pre–Renaissance Europe, when it was actually the extension of its very opposite. Instead of promoting a present-based, peer-to-peer economy, the spectacle was reinforcing the time-delayed, long-distance economics of international monopolies.

For the first time, people engaged with products completely divorced from the people who actually made them. Technologies masked not just the labor, but also the time that went into an item's production. People walking through the Crystal Palace were living in a different temporal reality than those working the looms of Delhi. This new way of interacting with things defined a new human identity for the very first time—that of the consumer.

The consumer lives in a present made possible by the temporal compression of others. He can consume something in minutes that may have taken months to manufacture and transport. While he may have a job himself during which the clock ticks normally, when he is in the store he is in a different time zone altogether, one that leverages the time of production into a frozen present of shopping. This new relationship between people and products no longer depended exclusively on international trade. Items manufactured three thousand miles across the country embodied the same invisible labor and time as those made overseas.

The more companies excelled at mechanized mass production, however, the more consumption needed to be done. Supply easily outstripped demand. By the early twentieth century, decorators such as Frank Baum (later the author of *The Wizard of Oz*) were being employed to stoke consumer desire with store displays. Baum was responsible for what we now think of as the store window, where an entire aspirational lifestyle would be represented in a single, frozen scene. Baum also devised the notion of using themes to denote different departments of the store. In the employ of John Wanamaker, Baum developed the first true bridal department, a stage set through which he defined an entire wedding aesthetic. Once a bride-to-be had ventured into this world, the only way to feel like a true participant was to purchase every item in the ensemble. Anything less would feel incomplete.

What Baum and his counterparts at other department stores realized was that they needed to tell stories and to create feelings instantaneously. Remember, even though the items for sale may have taken time to produce, the consumer exists in no-time. In a store window, there is no linear time in which to convey a narrative. Instead, store windows must communicate through a still image—like the Nativity scene in front of a church at Christmas. Similarly, the individual departments pushed customers to act by making them feel out of place at that very moment. After World War II, these efforts focused almost entirely on women, who were being counted on to consume enough to keep returning veterans at the plants, churning out merchandise. By the 1950s, the need to consume and participate in these fantasy worlds became so highly pressurized that women began stealing compulsively—a condition called kleptomania, which was finally recognized by the American Psychiatric Association in the early 1960s. (Interestingly, women were the only identified sufferers of the condition in the 1950s and '60s. However, after women went back to work in the 1970s and marketers turned their attention to children, teenagers became the primary sufferers of kleptomania in America.) Kleptomania is just a symptom of this chronically induced inferiority—another way of expressing overwind. In the short forever of retail shopping, we are always out of step, hoping to buy our way into these worlds.

Mass media reinforced the perpetual and insatiable sense of longing, presenting consumers with imaginary landscapes of happiness. While mass production may have disconnected the worker from the value of his skill set, mass marketing disconnected the consumer from the producer. Instead of buying products from the person who made them, consumers were now purchasing them from companies thousands of miles away. In order to re-create the real-time human relationships that people had with their local millers, druggists, and butchers, long-distance sellers developed brands. The face of the Quaker on an otherwise plain box was meant to instill the sense of kinship one used to feel with the miller.

Mass media arose, at least in part, to forge that relationship in advance. Advertisements in print and commercials on television feed us the mythology of a brand so that it is spring-loaded into our psyche—ready to emerge fully formed when we see the label in the store. Our shopping experience—just like the experience of the attendees at the Crystal Palace—is no longer occurring on a normally flowing timeline, but is instead a series of decompressions. Each label we see recalls and unpacks advertisements and commercials, which in turn unpack the cultural mythologies they have appropriated.

Of course, the consumer must never be allowed to reach his goal, for then his consumption would cease. The consumer must never feel completely at home in his present, or he will stop striving toward a more fully satisfied future. Since consumption makes up about half of all economic activity in America, a happy consumer would spell disaster. Fashion must change, and products must be upgraded and updated. In order for the economy to grow, this must keep happening faster.

The economics of consumption have always been dependent on illusions of increasing immediacy and newness, and an actuality of getting people to produce and consume more stuff, more rapidly, with evermore of their time. The expectations for instant reward and satisfaction have been built up by media for close to a century now. The amount of time between purchase (or even earning) and gratification has shrunk to nothing—so much so that the purchase itself is more rewarding than consuming whatever it is that has been bought. After waiting several days in the street, Apple customers exit the store waving their newly purchased i-gadgets in the air, as if acquisition itself were the reward. The purchase feels embedded with historicity. They were part of a real moment, a specific date. The same way someone may tell us he was at the Beatles famous concert at Shea Stadium, the Apple consumer can say he scored the new iPhone on the day it was released.

Where "act now" once meant that a particular sales price would soon expire, today it simply means there's an opportunity to do something at a particular moment. To be a part of something. Or simply, to *do* something. Just as midcentury mass-produced goods now feel historical and authentic to us, there's a sense that reaching the crest of current innovation or style will keep us authentic to the now. It's not the reality of a bridal display that keeps evading us so much as the standing wave of novelty and variation. Even if the *now* is standing still, what we require to be fully equipped for this moment keeps changing.

"Act now" becomes less a call to action than an ongoing call for activity. In a digitally enhanced consumer reality, we not only work to keep up with the latest products and service options, we purchase products and services that serve no purpose other than to help us better keep up. Our iPads and Androids are nothing like the productivity-computing tools on which they may once have been based but are instead purchasing platforms designed to increase the ease and speed with which we consume.

Speeding up all this consumption further is the fact that we don't even have to take ownership of the things we consume. New incarnations of the things we buy are successively less tangible, reducing the friction associated with purchasing, using, and disposing of real objects. Instead of buying cars, we lease them. Instead of purchasing homes, we sign mortgages. As if coming full circle to the era of lords and vassals, we no longer own land at all but simply pay for the right to use it.

Likewise, music ownership used to mean purchasing a physical vinyl disk—an artifact of a live performance. Then we replaced our vinyl collections with digitally stored music on CDs. Then we began purchasing music by downloading files from the net, with no physical medium at all. And now we are purchasing subscriptions to services such as Spotify or Turntable.FM through which we acquire the right to listen to music we never own in any form. Our right to use is in the present tense only; it is a constantly expiring resource requiring ongoing replenishment.

And oddly enough, the less tangible and more digital our music and other consumption becomes, the less sharable and open source it is rendered. The record can be taped or sold, and the CD can be copied and shared online. The downloaded, copy-protected file is personally owned on one's own personal device, with encrypted prohibitions on sharing. This is also true for literature. Books, once passed from person to person when finished, or left in a hotel room for the next guest to find, are now read as e-books, tied to a single user and device.

In this final iteration of spring-loading, consumption becomes more like attending a performance. The consumer is no longer truly consuming anything, but experiencing and paying for a constant flow of user rights to things, services, and data owned by others. In one sense, this is freeing. There's no need to organize or back up precious digital files for fear of losing them, no books and records to box up every time one moves. Less stuff to get damaged in a fire or flood, fewer resources depleted, and not as many objects to throw away. But it is a reality in which contracts define our access, corporations invade our privacy, and software limits our ability to socialize and share.

We get so much better and faster at consuming all the time that there's no point in actually *having* anything at all. In a certain light, it sounds almost communal. Except we are not building a new commons together where everything is shared; we are turning life into a set of monetizable experiences where the meter is always on.

TIME IS MONEY

One would think all this innovation in consumption should at the very least yield greater profits for the corporations stoking it. It turns out this is not the case. According to the financial analysts at Deloitte & Touche, asset profitability for US firms has steadily fallen 75 percent over the past forty years.[21] While corporations have successfully accumulated most of the financial resources out there, they don't know how to reinvest it to make more.

So no matter how much or rapidly we humans consume, no matter how many mental conditions we are willing to acquire on consumption's behalf, we seem incapable of making this work for business on a financial level. We just can't go any faster. Even when we consume and dispose of resources at a pace that threatens the ability of our environment to sustain human life, we can't consume rapidly enough to meet the demands of the market for growth.

In America, certainly, there is already more than enough stuff to go around. We have constructed so many houses that banks are busy tearing down foreclosed homes in order to keep market value high on the rest of them. The US Department of Agriculture burns tons of crops each year in order to prevent a food glut that will impact commodity prices. Viewed in this light, our challenge with unemployment is less a problem of an underskilled population than that of an overskilled one—or at least an overproductive one. We are so good at making stuff and providing services that we no longer require all of us to do it. As we are confronted by bounty, our main reason to create jobs is merely to have some justification for distributing all the stuff that is actually in abundance. Failing that, we simply deny what is available to those in need, on principle.

We cannot consume ourselves out of this hole, no matter how hard we try, and no matter how much time we compress into each consumptive act. This is because we are asking our consumption to compensate for a deeper form of time compression—one built into the landscape of economics itself. For not only is time money, but money is time.

We tend to think of money as a way of stopping time: As psychologist Ernest Becker argued in his classic text *The Denial of Death*, our bank accounts are emotional stand-ins for survival. We accumulate money as a substitute for being able to accumulate time. The time we have left is always an unknown; the money we have left is quite certain. It is solid—or at least it once was. Gold held its value over time, no matter who was in charge, what the weather did, or which side won the war. Money was valued for its durability and solidity. This was especially true after local coinage was outlawed in favor of long-distance currencies. People had readily accepted the value of their local currencies, even though they were printed on worthless foil, because they personally knew the grain stores accountable for them. Centrally issued currencies were more impersonal and had to function across much wider distances. Monarchs were not implicitly trusted, nor were their reigns guaranteed, so they were forced to include a standardized measure of scarce metal into their coins for them to be accepted.

In spite of this outward bias toward storage and solidity, centrally issued currency actually had the effect of winding up a nation's economic mainspring. That's the impact of simple interest on money: interest-bearing currency isn't really just money; it is money over *time*.

Money used to grow on trees—or, rather, out of the ground. Local currencies were earned or, quite literally, grown into existence by grain farmers. Cash was as abundant as the season's harvest, and its relative value fluctuated with the size of the crop. This wasn't really a problem, because the purpose of money was to allow for transactions. As long as people understood what their money was worth, they could use it.

Central currency is loaned into existence, at interest. Most simply, a person who wants to start a business borrows $100,000 from the bank, with the requirement that he pay back, say, $200,000 over the next ten years. He has a decade to double his money. Where does the additional $100,000 come from? Ultimately, from other people and businesses who are in the same position, spending money that they have borrowed. Even the wages that workers receive to buy things with were borrowed somewhere up the chain.

But this seems to suggest a zero-sum game. Each borrower must win some *other* borrower's money in order to pay back the bank. If the bank has loaned out $100,000 to ten different businesses, all competing to earn the money they need to pay back their loans, then at least half of them have to fail. Unless, of course, someone simply borrows *more* money from the bank, by proposing an additional business or expansion.

Therein lies the beauty and horror of interest-bearing currency. Interest is expansionary. As long as the economy is growing, everything works out. The requirement to pay back at the rate of interest motivates businesses as surely as the loan shark encourages his borrowers to keep up their weekly installments. Running a business and growing a business end up meaning the same thing. Even if one business pays back everything it owes, this only puts some other business into debt. As the debtor seeks to expand to meet its interest requirements, the debtor either takes territory from an existing business or finds a new territory. Standing still is to lose.

That's why in the centuries following the implementation of interest-bearing currency, we saw such rapid and, in many cases, merciless expansion of colonial European powers across the globe. They had no choice. The bias of the money supply toward growth biased these powers toward growth, too. Interestingly, the Ottoman Empire utilized a series of noninterest-bearing regional currencies under the millet system and did not suffer the same growth requirement. While the empire still had its conquests, they were not economically required for the fiscal system to remain solvent. Sustainability was still an option.

To be sure, in the six centuries following the birth of modern-style centrally issued currencies, we witnessed expansion and advancement on an unprecedented scale. Europe colonized China, India, Africa, the Middle East, the Americas, and islands in every ocean. By the twentieth century, the United States had joined the territorial race as well. (Only France, with its limited navy, proved incapable of expanding in quite the same way, leading finance minister Jean-Baptiste Colbert to invent the concept of French luxury and achieve expansion through exports. "French fashions must be France's answers to Spain's gold mines in Peru," he declared.)[22]

Once the overt conquests of nations and the subjugation of their people was no longer feasible, the West achieved the same thing through more virtual means. After World War II, the last of the European colonies—such as India and Palestine—were proving ungovernable. The creation of the World Bank and the International Monetary Fund gave Western powers a new way to expand their economies without actually taking over countries. Instead, in the name of liberating these regions, they would lend large sums to so-called developing nations, at interest.

In return for the privilege of going into debt, the borrowing nations would also be required to open themselves to unrestricted trade with lending nations' corporations. This meant foreign companies were entitled to build factories, undercut local industry or farming, and otherwise turn these fledgling nations into colonies—now called beneficiaries. Open-market trading rules written by the bank invariably favored the foreign power. Indigenous populations grew poorer. They were incapable of competing with the corporations that set up shop and often lost access to subsistence farming due to both disadvantageous trade policies and the pollution and runoff from foreign factories. Amazingly, toxic spills and skyrocketing health issues were generally recorded as increases to the nation's GNP and touted in World Bank literature as success stories. Violence and value judgments aside, the debt requirement was still being served, and economic expansion was allowed to continue.

A similar strategy was employed on Western consumers. The emergence of credit cards, federally preapproved residential mortgages, and other lines of credit in the mid-twentieth century served a growth economy in two ways. First, and most simply, it gave people a way to buy more stuff now than they actually had money to buy. This is pure time compression: the money a person is likely to earn in the future is packed into the present moment. Second, the creation of an indebted, interest-paying population expands the economy. Instead of just paying three thousand dollars for a new car, the consumer spends nine thousand dollars by the time interest payments are done. That nine thousand dollars goes on the books as nine thousand of projected revenue. Easy growth. The way for businesses to service their own debt became simply to pass that debt on to their customers, ideally at higher rates of interest than they paid for the money themselves.

And just as the World Bank helped struggling nations envision promising futures as participants in the First World economy, retail banks and credit companies exploited well-known temporal distortions to convince people how easy it would be to pay back loans in the future. The practice of behavioral finance seeks to exploit both the loopholes in lending rules as well as the documented lapses in people's understanding of the consequences of their own decisions. It is not a fringe science. JPMorgan Chase has an entire U.S. Behavioral Finance Group that is dedicated to looking at investor emotions and irrationality, particularly during crises.[23] Behavioral economists now teach at Stanford, Berkeley, Chicago, Columbia, Princeton, MIT, and Harvard,[24] and have earned two Nobel Prizes in economics.

Behavioral finance is the study of the way people consistently act against their own best financial interests, as well as how to exploit these psychological weaknesses when peddling questionable securities and products. These are proven behaviors with industry-accepted names like "money illusion bias," "loss aversion theory," "irrationality bias," and "time discounting." For instance, people do not borrow opportunistically, but irrationally. As if looking at objects in the distance, they see future payments as smaller than ones in the present—even if they are actually larger. They are more reluctant to lose a small amount of money than gain a larger one, no matter the probability of either event in a particular transaction. They do not consider the possibility of any unexpected negative development arising between the day they purchase something and the day they will ultimately have to pay for it. Present shock.

Banks craft credit card and mortgage promotions that take advantage of these inaccurate perceptions and irrational behaviors. Zero-percent introductory fees effectively camouflage regular interest rates up to 30 percent. Lowering minimum-payment requirements from the standard 5 percent to 2 or 3 percent of the outstanding balance looks attractive to borrowers.[25] The monthly payment is lower! However, the additional amount of time required to pay off the debt at this rate will end up costing them more than *triple* the original balance. It is irrational for them to make purchases and borrow money under these terms, or to prefer them to the original ones.

It proved just as irrational for American banks to depend on this domestic and international sleight of hand for so long. Eventually debtor nations began either defaulting on their loans (Argentina) or insisting on being included in the value chain (China). Lender nations soon found themselves in debt and nursing negative trade balances. Right around the same time, American consumers began to realize they could not support their own rate of consumption, even when leveraged by credit. Expansion seemed to be threatened, as the available surface area for new transactions had proved limited. Where to grow?

Instead of innovating in the real world, banks turned to the financial instruments themselves. Interest had worked just fine for centuries, squeezing wealth out of money itself. Could the formula be repeated? Could more time be packed into interest?

Interest had given money the ability to generate wealth through lending. Vastly simplified, interest is just a way of expressing money over time. Lend some money and over time that money multiplies. Whether it is a bank lending money to a business, an investor buying some shares of stock, or a retiree buying a bond, everyone expects a return on his investment over time.

As these first-order investments began to slow down—at least compared with the rate at which the economy needed to grow—everyone began looking for a way to goose them up. That's where derivatives came in. Instead of buying shares of stock or whole bonds or mortgage notes, derivatives let investors bet directly on the changing value of those instruments over time. Instead of buying actual stocks and bonds, investors buy the right to buy or sell these instruments at some point in the future.

Most of us understand how derivatives let investors make bigger bets with less money up front. Investors can purchase the options on a stock for a tiny fraction of the cost of the stock itself. They only pay for the actual stock once they exercise the option. Hopefully, the shares are worth a lot more than the strike price on the option, in which case the investor has made money.

But on a subtler level, the investor hasn't merely leveraged his money; he has leveraged time. The option is a financial spring-loading mechanism, packing the future's price fluctuations into today's transactions. The investor is no longer investing in a company or even its debt, but rather in the changing value of that debt. In other words, instead of riding up the ascending curve of value, derivatives open the door to betting on the rate of change. Value over time, over time.

In the effort to compress time ever further, this process can be repeated almost ad infinitum. Traders can bet on the future price of derivatives—derivatives of derivatives—or the future price of those, or even just the volatility of price swings. At each step along the way, the thing being invested in gets more abstract, more leveraged, and more time compressed. In the real world, value can't be created fast enough to keep up with the rate of expansion required by interest-driven currency, so it instead gets compressed into financial instruments that pack future value into present transactions.

Regular people end up compressing time into money the same way. Back when I still believed I could afford to purchase a three-bedroom apartment in New York City, the real estate agent showed me residences far out of my price range. She explained to me that I could qualify for an ARM, or an adjustable-rate mortgage, with a special introductory interest-only rate for the first five years. This way, I would have a monthly payment I could afford, even though I wouldn't be working toward reducing the principal. After the first five years, the loan would adjust to a more normal rate—one much greater than I could afford. She told me that wouldn't be a problem, since at that point I could simply refinance with another five-year ARM.

The idea was that that apartment would be worth more by the time I needed to get a new mortgage, so I would be able to refinance at a higher assessed value. As the market kept going up, the meager portion of the apartment I owned would be going up in value as well, giving me the collateral I would need to refinance. Each time, the cash amount I need to finance remains the same, but the percent of the apartment that debt represents goes down. So, in theory, the more the price of the apartment goes up, the more I own, and the more easily I can refinance.

Of course, that didn't happen. Luckily for me I didn't buy the apartment (or, rather, I didn't buy the mortgage for that apartment). But the hundreds of thousands of Americans who did accept similar bargains ended up in big trouble. Instead of increasing, the value of the homes sunk below the amount that was owed on them. As of this writing, 31 percent of all residential properties with mortgages are under water, or what industry analysts call negative equity.[26] Owing more on a thirty-year mortgage than one's house is currently worth is just another way of saying present shock.

A few traders did see the writing on the wall and understood that the housing market had become too dependent on these temporally compressed lending instruments. Famously, even though they were selling packaged loans to investors and pension funds, Goldman Sachs determined that the financing craze was unsustainable and began betting against the mortgages through even more derivative derivatives called credit default swaps. When it came time for the company on the other side of those default swaps, AIG, to pay up, only the US government could print enough money to bail them out.[27]

By today's standards, however, Goldman's successively derivative bets seem almost quaint. Their investing decisions were still based in what they saw as the likely future of an unsustainable system. Their contribution to the tragedy from which they hoped to benefit notwithstanding, they were making a prediction and making a bet based on their analysis of where the future was heading. Temporally compressed though it may be, it is still based on making conclusions. Value is created over time. It is a product of the cause-and-effect, temporal universe—however much it may be abstracted.

A majority of equity trading today is designed to circumvent that universe of time-generated value altogether. Computer-driven or algorithmic trading, as it is now called, has its origins in the arms race. Mathematicians spent decades trying to figure out a way to evade radar. They finally developed stealth technology, which really just works by using electric fields to make a big thing—like a plane—appear to be many little things. Then, in 1999, an F-117 using stealth was shot down over Serbia. It seems some Hungarian mathematicians had figured out that instead of looking for objects in the sky, the antiaircraft detection systems needed to look only for the electrical fields.[28]

Those same mathematicians and their successors are now being employed by Wall Street firms to hide from and predict one another's movements. When a bank wants to move a big quantity of shares, for example, it doesn't want everyone to know what it is doing. If news of a big buy leaked out before the big buy could be completed, the price may go up. To hide their motions, they employ the same technique as stealth planes: they use algorithms to break their giant trade into thousands of little ones, and do so in such a way that they look random. Their sizes and timing are scattered.

In order to identify this stealthy algorithmic movement, competing banks hire other mathematicians to write other algorithms that monitor trading and look for clues of these bigger trades and trends. The algorithms actually shoot out little trades, much like radar, in order to measure the response of the market and then infer if there are any big movements going on. The original algorithms are, in turn, on the lookout for these little probes and attempt to run additional countermoves and fakes. This algorithmic dance—what is known as black box trading—accounts for over 70 percent of Wall Street trading activity today.

In high-frequency, algorithmic trading, speed is everything. Algorithms need to know what is happening and make their moves before their enemy algorithms can react and adjust. No matter how well they write their programs, and no matter how powerful the computers they use, the most important factor in bringing algorithms up to speed is a better physical location on the network. The physical distance of a brokerage house's computers to the computers executing the trades makes a difference in how fast the algorithm can read and respond to market activity.

As former game designer Kevin Slavin has pointed out in his talks and articles,[29] while we may think of the Internet as a distributed and nonlocal phenomenon, you can be closer or farther from it depending on how much cable there is between you and its biggest nodes. In New York, this mother node is fittingly located at the old Western Union Building on 60 Hudson Street. All the main Internet trunks come up through this building, known as a colocation center, or carrier hotel. Its fiber optic lines all come together in a single, 15,000-square-foot "meet me" room on the ninth floor, powered by a 10,000-amp DC power plant.

If a firm owns space anywhere in that building, its computers are sitting right on the node, so its algorithms are operating with a latency of effectively zero. Algorithms running on computers all the way down on Wall Street, on the other hand, are almost a mile away. That adds up to about a two-millisecond delay each way, which means everything. A fast computer sitting in the carrier hotel can see the bids of other algorithms and then act on them before they have even gone through. The purpose of being in close isn't simply to front-run the trade but to have the ability to fake out and misdirect the other side. Algorithms don't care what anything is really worth as an investment, remember. They care only about the trade in the present.

In a virtual world of black box trading where timing is everything, getting closer to the "meet me" room on the ninth floor of 60 Hudson Street is worth a lot of money. Firms are competing so hard to position their computers advantageously that the real estate market in the neighborhood has been spiking—quite unpredictably, and only explained by the needs of these algorithms for quicker access. Architects are busy replacing the floors of buildings with steel in order to accommodate rooms filled with heavy servers. Both the real estate market and the physical design of Lower Manhattan are being optimized for algorithms competing to compress money into milliseconds. It is becoming a giant microchip or, better, a digital stopwatch.

When the only value left is time, the world becomes a clock.

LIVING IN RAM

When an economy—or any system, for that matter—becomes so tightly compressed and abstracted that only a computer program can navigate it, we're all in for some surprises. To be sure, algorithms are great at increasing our efficiency and even making the world a more convenient place—even if we don't know quite how they work. Thanks to algorithms, many elevators today don't have panels with buttons. Riders use consoles in the lobby to select their floors, and an algorithm directs them to the appropriate elevator, minimizing everyone's trip time. Algorithms determine the songs playing on Clear Channel stations, the ideal partners on dating websites, the best driving routes, and even the plot twists for Hollywood screenplays—all by compressing the data of experience along with the permutations of possibility.

But the results aren't always smooth and predictable. A stock market driven by algorithms is all fine and well until the market inexplicably loses 1,000 points in a minute thanks to what is now called a flash crash. The algorithms all feeding back to and off one another get caught in a loop, and all of a sudden Accenture is trading at $100,000 a share or Proctor & Gamble goes down to a penny.[30] Ironically, and in a perfect expression of present shock, the leading high-frequency trading exchange had a high-profile flash crash on the same day it was attempting to conduct its own initial public offering—and on the same day I was finishing this section of the book. The company, BATS Global Markets, runs a stock exchange called Better Alternative Trading System, which was built specifically to accommodate high-frequency trading and handles over 11 percent of US equities transactions. Their IPO was highly anticipated and represented another step in technology's colonization of the stock exchange.

BATS issued 6 million shares for about $17 each, and then something went terribly wrong: their system suddenly began executing trades of BATS stock at three and four cents per share.[31] Then shares of Apple trading on the BATS exchange suddenly dropped 10 percent, at which point the company halted trading of both ticker symbols. Embarrassed, and incapable of figuring out quite what happened, BATS took the extremely unusual step of canceling its own IPO and giving everyone back their money.

What will be the equivalent of a flash crash in other highly compressed arenas where algorithms rule? What does a flash crash in online dating or Facebook friendships look like? What about in criminal enforcement and deterrence, particularly when no one knows how the algorithms have chosen to accomplish their tasks? Algorithmic present shock is instantaneous. Its results impact us before it is even noticed.

At least on the stock market, participation in this pressure cooker is optional. Many investors are no longer willing to take the chance and have decided not to share their marketplace with algorithms that can outcompress them or, worse, spin the whole system out of control. Stock traders are leaving Wall Street in droves and opting for the dark pools of Geneva where they can exchange shares anonymously. While dark pools may have first served simply to conceal institutional-trading activity from the public, now they are being used by investors who want to conceal their trading activity from algorithms—or simply distance themselves from the effects of algorithmic spring-loading.[32]

They have resorted to one of the two main strategies for contending with the perils of overwinding in the short forever, which is simply to avoid spring-loaded situations altogether. Traders lose the advantages of superconcentrated, high-frequency trades, but regain access to the accretive value of their investments over time. They lose the chance to cash in on the volatility of the marketplace irrespective of what things may be worth, but get to apply at least a bit of real-world knowledge and logic to their investment decisions. Plus, they are insulated from the chaotic feedback loops of algorithms gone wild.

It's not a matter of abandoning the present in order to make smart investment decisions and more secure trades. If anything, nonalgorithmic trading represents a return to the genuine present of value. What is this piece of paper actually worth right now in terms of assets, productivity, and, yes, potential? Rather than packing in nine derivative layers' worth of time over time over time into a single inscrutably abstract ticker symbol, the investment represents some comprehensibly present-tense value proposition.

There are many other healthy examples of businesses, communities, and individuals that have opted out of spring-loading in order to enjoy a more evenly paced world. In Europe, where the leveraging of the long-distance, debt-generating euro currency has proved catastrophic for local industry and commerce, people are turning to alternative currencies that insulate them from the macroeconomic storm around them.

In the Greek town of Volos, for example, citizens are experimenting with a favor bank. Mislabeled a barter economy by most journalists, this effort at a cashless economy is actually an exchange network.[33] Everyone has an account online that keeps track of how many Local Alternative Units, or tems, they have earned or spent. People offer goods and services to one another, agree to terms, and make the exchange. Then their accounts are credited and debited accordingly. No one is allowed to accumulate more than 1,200 tems, because the object of the system is not to get individuals wealthy by storing currency over time, but to make the community prosperous by encouraging trade, production, and services.

The man who devised the website for one of these networks says that the euro crisis had rendered the Greeks "frozen, in a state of fear. It's like they've been hit over the head with a brick; they're dizzy. And they're cautious; they're still thinking: 'I need euros, how am I going to pay my bills?'"[34] This is the vertigo of the short forever, present shock caused by absolute dependence on a time-bound medium of exchange. It was so overleveraged that when it broke, it exploded like an overwound spring. The scales of interaction were an almost quantum mismatch: global investment banks and national treasuries do not have the same priorities as tiny villages of people attempting to provide one another the foods and basic services they need to sustain themselves. The macroeconomics of Greek debt—as well as that of the Spanish, Italian, and Portuguese—accumulated and compressed over decades, but unraveled in an instant, overwhelming the real-time economy of people.

The scale and nature of their new form of exchange encourages real-world, real-time interaction as well. Members of the various local networks meet on regular market days to purchase goods or negotiate simple service agreements for the coming week. This is a presentist economy, at least in comparison with the storage-based economics of the euro and traditional banking. Nothing is spring-loaded or leveraged, which makes it harder for these markets to endure changing seasonal conditions, support multiyear contracts, or provide opportunities for passive investment. But this style of transaction still does offer some long-term benefits to the communities who use it. Human relationships are strengthened, local businesses enjoy advantages over larger foreign corporations, and investment of time and energy is spent on meeting the needs of the community itself.

Greek villagers in the shadow of a failing euro aren't the only ones abandoning time-compressed investment strategies. The town of Volos represents just one of hundreds of similar efforts throughout Europe, Asia, and Africa, where centrally administrated, interest-bearing currencies no longer support the real-time transactions of people—or ask so much for the privilege as to be impractical for the job. These currencies are not limited to struggling, indebted nations but are emerging in the lending economies such as Germany as well, where macrofiscal trends inhibit local transactions.[35]

Meanwhile, on the other side of the economy—and the world—commodity traders have been present shocked by the real-time nature of today's agriculture and marketplace. Commodity traders generally buy and sell futures on particular goods. Their function (other than making money for themselves) is to help farmers and buyers lock in prices for their goods in advance. A clothing manufacturer, fearing rising cotton prices, may want to purchase its crop in advance. The commodity trader is hired to purchase futures on cotton, specifying the date and price. Meanwhile, a cotton grower—who suspects that prices on his commodity may actually go down over the next year—may want to get cash at today's rates. He calls his commodity trader, who makes a deal across the commodities pit with the trader working for the clothing maker.

The traders themselves understand the subtleties of the time compression that they are performing for their clients. Their business is about storing value, wagering on the future, and then using contracts to leverage present expectations against future realities. They may make a lot of money without creating any tangible value, but they do help create liquidity in markets that need it and force a bit of planning or austerity when something is going to be in short supply.

However, new technologies and global supply chains are turning formerly seasonal commodities into year-round products. And consumers' decreasing awareness of seasons is changing what we expect to find at our local market and when. This longer now of both supply and demand cycles is doing to many commodities traders what twenty-four-hour cable did to the evening news: in a world of constant flow, the ability to compress time becomes superfluous.

The pork belly commodity pits fell to this form of present shock just a few years ago. The pork belly trade—conducted primarily in a trading pit at the Chicago Mercantile Exchange—allowed investors to bet on the ebbs and flows of supply and demand. Hogs were slaughtered in certain seasons, and bacon was consumed most heavily in others. Commodity traders served to spring-load and unwind both sides of the trade. They bought up shares of frozen bellies and stored them in Chicago warehouses, and then sold them off when demand rose in March and April for Easter brunch and then again in late summer for the tomato harvest and everyone's BLT sandwiches. The traders created storage when there was only flow, and flow when there was only storage. Pork bellies became the most popular product on the Chicago trading floor.

Consumers no longer understand why they should settle for frozen bacon during certain parts of the year, and pork suppliers have learned how to accommodate the demand for constant flow. With no need for their ability to translate flow into storage and back again, commodity traders left the pork bellies pit, and the oldest existing livestock futures contract was delisted. Futures traders became disintermediated by a marketplace looking for direct access to an always-on reality.

So while local merchants in depressed European economies are abandoning storage-based euros because it is too expensive for them to use, the pork industry stopped using futures because their focus on freshness and flow was no longer compatible with a spring-loaded strategy. Still other businesses are learning that moving from storage to flow helps them reduce their overleveraged dependence on the past and to take advantage of feedback occurring in the moment.

Walt Disney's theme parks, for just one great example, began losing their luster and profit margins in the mid-1990s. Disney's CEO at the time, Michael Eisner, wasn't quite sure what was going wrong. He had been milking Walt Disney's treasure trove of icons and intellectual property quite successfully in media for some time, but back in the real world, competition from Six Flags, Universal Studios, and other theme parks appeared to be moving in on Disney's iconic theme park business. Walt Disney had died in 1966, long before most of the current staff at the parks had been born. Perhaps a new chief of Disney's theme parks company could revive the spirit of Disney and teach the park's people how to exude more of Walt's trademark magic?

Eisner tapped his CFO, Judson Green, to put the parks on a growth track consonant with the greater studio's expansion into radio, television, and sports. Could Green release the stored-up potential of Mickey, Donald, and the rest of Disney's fabled intellectual property? Green took a completely different tack. Instead of relying on Disney's spring-loaded century of content, he realized that a theme park always exists in the present. The key to reviving its sagging spirit was not to disinter and channel Uncle Walt, but to break the parks' absolute dependence on the stored genius and brand value of the legendary animator.

Instead, Green turned the theme parks into an entirely presentist operation. This had less to do with updating rides or rebuilding Tomorrowland (both of which he did) than it did with changing the way communication and innovation occurred on the ground. Green's philosophy of business was that innovation doesn't come from the top or even from stored past experience, but from people working in the moment on the front lines. Management's job is not to fill current employees with the collected, compressed wisdom of the ages, but rather to support them in the jobs only they are actually charged with doing. Management becomes a bit like a customer service department for the employees, who are the ones responsible for the business.

For example, a few years ago Disneyworld guests began complaining that they couldn't fit their rented strollers onto the little railroad train that goes around the park. Traditionally, a company might look to solve the problem from the top down by, say, redesigning the trains. But the company tasked its employees with finding a solution. And sure enough, an employee came upon the obvious fix: let the guest turn his stroller in at one train location and then pick up a new one at the next. Removable stroller name cards were created, and the problem was solved. To the Disney guest, Walt's magic was alive and well in the continual flow of solutions, and every service metric went through the roof along with efficiency and revenue. Green's management style differed from Eisner's, however, and the executive who transformed the parks from historical landmarks to innovation laboratories was terminated.[36]

Understandably, Michael Eisner was nervous about letting go of his company's ample hard drive and living completely in RAM. After all, he had begun his tenure at the helm by reviving the *Wonderful World of Disney* television show on Sunday evenings and then introducing the show on camera, in the style of Walt himself. He also licensed Disney imagery, seemingly everywhere, and traded on the Disney name and brand to make business deals that may not have been entirely in keeping with Disney's tradition of putting quality over profit. His impulse to squander so much of Disney's stored legacy all at once did boost Disney's stock price but horrified Disney's heirs. Roy E. Disney claimed Eisner had turned Disney into a "rapacious, soul-less" company, resigned from the board, and began the process that led to Eisner's own resignation a few years later.[37] In his own book on the period, Eisner still seems oblivious to the fundamental issue at play, preferring to see it as a clash of management styles and personalities than a culturewide case of present shock.

Balancing the needs of the present against the stored leverage of the past is a tricky proposition, and where so many of us get confused. How many generations before our own asked people to earn and save enough money while young in order to accumulate a nest egg big enough to live off for the last third of one's life? It's a bit like asking an animal to get fat enough not to have to eat again for the rest of its natural life.

But the opposite (however attractive from a survivalist's perspective) doesn't really work, either. Living completely in RAM is a bit like living on an Atkins diet. Everything just goes through you. With no bread, no starch, there's nothing to hang on to. There's no storage, no accumulation. There's never any leverage, just constant motion—all flow. An absolutely presentist approach to business, or life for that matter, is untenable. Except for a Buddhist monk, literally walking the streets with nothing but a bowl in hand, who can live this way much less run a business or organization? No one.

Instead, we learn to spring-load appropriately. We compress and unpack time in ways that support the present moment without robbing from the future or depleting our reserves. We don't pack so much in that we hoard and weigh down the present, but we don't get so lean that we cannot survive without uninterrupted access to new flow. We learn to spring-load without overwinding.

WINDING UP

Instead of avoiding all time compression whatsoever, we can use our knowledge of its strengths and weaknesses to our advantage. There are ways to pack time into projects in advance, so that the vast hours or years of preparation unfold in the instant they are needed.

As the first fire-using cave people realized, the natural world is spring-loaded with energy that can be released in the present as heat. A tree is a few hundred years of sun energy, stored in wood. Oil stores still more years of energy, and coal even more. That's why they're called *fossil* fuels. We release the densely packed investment of millennia of life in order to power our world right now. In doing so, we deplete the reserves available for the future faster than they can be replenished. We also pollute the future faster than it can be cleaned.

When storage fails, we turn to flow. In the face of fossil fuel's shortcomings, we are attracted to sustainable and renewable energy resources such as wind and solar. They don't appear to deplete any-thing: the sun keeps shining and wind keeps blowing. They don't leave any residue for our grandchildren to breathe in or clean up. The problem is that they are not truly continuous. The sun goes away at night, and winds are irregular. The answer, again, is to compress and store the energy of one moment to use at another.

This is harder than it seems. The current barrier to renewable energy (other than reluctance to disrupt today's fossil fuel markets with a free alternative) is our inability to store and distribute energy efficiently. Current battery technology is not only inadequate to the job, it also requires the mining of rare earth metals such as lithium and molybdenum, which are as easy to deplete as oil. We end up weaning our cars off oil only to make them dependent on other even more leveraged and scarce battery materials. The solution lies in our ability to find ways to compress and unwind energy in as close to real time as possible, so that the materials we require for battery storage needn't be as concentrated as lithium. It also in-volves learning to spend our energy more frugally than we did in the twentieth century, when we had no idea coal and oil—or that atmosphere—may be in limited supply. Nature stores energy on a very different timescale than human combustion technologies can burn it.

Maybe this is why Prometheus was punished by the gods for bringing fire to mortals. Perhaps the tellers of this myth sensed that fire was unwinding the stored time of the gods—of nature—and unleashing a force beyond human understanding or control. Inter-estingly, his punishment was a torturous present shock: he was tied to a cliff where an eagle consumed his liver, perpetually.

We humans cannot live entirely in the short forever, however. Building a civilization has meant learning to store time and spring-load things carefully and deliberately. We can do so without anger-ing the gods or destroying our world. Spring-loading doesn't even have to be an artifact of the fossil record; it can be an entirely intentional and predictable strategy.

Taking their cue from nature, many businesses and organizations have learned to pack time into one phase of their work so that it can spring out like a fully formed pup tent when it is needed. The Shaare Zadek Medical Center employed this strategy to erect an instant set of operating rooms, clinics, and wards in a soccer field in Japan, serving victims of the 2011 tsunami. Although field hospitals have been used by the military for close to a century now, the doctors at Shaare Zadek took this concept to a whole new level by creating ready-to-ship, expandable medical centers that can be air-dropped virtually anywhere. "If you drop our group in the middle of a desert, we can work," explains one of the hospital's cardiac surgeons.[38]

Shaare Zadek has been working on the process since 1979 but reached public attention after the Haitian earthquake of 2010. The Israeli hospital achieved mythic status, providing earthquake victims with a level of care that wasn't accessible to them even under normal circumstances, such as modern respirators and properly cross-matched blood transfusions. In Japan, Shaare Zadek was not only the first field hospital on the ground, it was also the most equipped, offering everything from obstetrics to ophthalmology. The field hospital has gone on twelve missions so far, each one serving as a new iteration that compounds upon the previous ones. As a time-management scheme, Shaare Zadek models spring-loading at its best: weeks of physical loading and preparation plus years of experience and learning are all packed into shipping containers that open and expand instantaneously on site, where they can be used in an emergency—when there is no time to spare.

We see new forms of spring-loading occurring across many different sectors of society, particularly as we move into increasingly digital spaces. Joichi Ito, Internet entrepreneur and director of

MIT's Media Lab, for example, believes that the development cycle of new technologies needs to be compressed. The formerly distinct processes of prototyping a product and releasing it onto the net have become the same thing. Arguing for flow over deliberation, Ito explains, "It is now usually cheaper to just try something than to sit around and try to figure out whether to try something. The product map is now often more complex and more expensive to create than trying to figure it out as you go. The compass has replaced the map."[39] In other words, instead of researching the market, prototyping a product, testing it, and then building a real version later, the developer compresses these processes into a single flow—planning, building, testing, and releasing simultaneously. It's the Tao of unwinding the past while winding the future, all in the same present.

Such a strategy depends on a community willing to help. Customers must see themselves as beneficiaries of their own investment in alpha- and beta-level software, competitors must see themselves as participants in the overall value creation unleashed by the project, and the initial developers must trust in their ongoing ability to innovate even after what they have shared is copied by others. It requires some underlying commons for it to work, because only the commons has a stake in the long-term repercussions of our actions. So far, most of us seem incapable of thinking this way.

That's what is meant by the tragedy of the commons. A bunch of individuals, acting independently and out of self-interest, may deplete a shared resource even though it hurts everyone in the long run. It applies to corporations that externalize costs such as pollution, but it's what happens when net users illegally download music and movies, expecting others to pick up the tab. It is in each person's short-term self-interest to steal the music. Only the sucker pays. But when everyone thinks that way, there's no one left to pay for the musician, and the music stops altogether.

The individualistic act of stealing the music or depleting the resource is a form of compression, robbing from the future to enjoy something in the present at no cost. As long as we live as individuals, the distant future doesn't really matter so much. The philosophy of the long now would suggest that the only way to see past this immediate, consumer-era self-satisfaction is to look further in the future. Have kids. Once we see that our long-term self-interest is no longer served, we may all, individually, change our behaviors. Even if we are thinking selfishly, prioritizing "me in the long run" isn't quite so bad as "me right now."

There's some good evidence that long-term thinking—even self-interested long-term thinking—pushes people to more collaborative solutions. In Prisoner's Dilemma experiments, for example, individuals are likely to testify against their accomplices in return for lighter sentences. The game is simple: two people are arrested and separated. If one testifies against his partner and the other remains silent, the betrayer goes free and the silent one goes to jail for a year. If both remain silent, they both get just one month in jail. If both betray, they each get three months.

So the worst possible outcome for an individual is to stay silent while the partner rats him out. This leads many to seek the safest, seemingly most self-interested way out—betray the opponent. The more times people play the game, however, the more likely they are to choose the more collaborative solution. They realize that they will have to play each other again and again, so it is now in their self-interest to demonstrate that they won't betray the opponent. The collaborating team gets the shortest total sentence if both stay silent. This sense of cooperation increases the greater the possibility of having to play again. As political scientist Robert Axelrod explains it, once "the shadow of the future"[40] lengthens, we have the basis for more durable relationships.

And that's certainly a step in the right direction. We are no longer content burning the past to pay for the present, because all of a sudden there's a future to worry about. I didn't truly start worrying about global warming or the depletion of water until I had a daughter, who by her very existence extended my mental timeline. But the more we ponder the future in this way, the more paralyzed we become by the prospect of the long now—frozen with that plastic bottle over the trash can.

Inconvenient truths tend to create more anxious neurotics than they do enlightened stakeholders. Those most successfully navigating the short forever seem to be the ones who learn to think wider, not longer. We must be able to expand our awareness beyond the zero-sum game of individual self-interest. It's not the longer time horizon that matters so much to alleviating our present shock as it is our awareness of all the other prisoners in the same dilemma. This is why the commons offers us not only the justification for transcending self-interested behavior but also the means to mitigating the anxiety of the short forever. The greater community becomes the way we bank our time and experience.

Or think of it this way: the individual is flow, and the community is storage. Only the individual can take actions. Only the community can absorb their impact over time. In presentist interactions, such as those conducted through alternative currencies in Greece, the individual can no longer hoard value. But the greater community is activated, and more goods and services move between people. The town that had no merchants, restaurants, or service providers now gets them, debt-free. Meanwhile, the community's experience with each vendor becomes part of its shared, stored, knowledge bank.

In a living community, one's reputation becomes the purest form of time binding and the easiest way of benefiting from the commons. The number of successful transactions next to your username on eBay is the community's accumulated experience with you as a seller and buyer, all boiled down into one number. In a network of self-interested individuals, such as the music stealers online, the files may be rated but not the people who uploaded them. They remain anonymous, not just because they are breaking the law, but because they understand that they are violating the commons.

The way to move beyond the paralyzing effects of the short forever is to stop trying to look so far into the individual futures of people or businesses, and instead to become more aware of what connects them to everyone and everything else right now.

CHAPTER 4

FRACTALNOIA

FINDING PATTERNS IN THE FEEDBACK

"Everything is everything," Cheryl declares, as if having solved the puzzle of life. The late-night-radio caller says she finally gets "how it all fits together."

She is speaking, at least initially, about chemtrails. Cheryl is convinced that the white vapor trails being drawn across the sky by planes actually contain more than jet exhaust. "The patterns have changed. And then when the tsunami hit, it all suddenly made sense. I could see the big picture."

The picture Cheryl has put together looks something like this: the condensation trails left in the wake of airplanes have been changing over the past decades. They form very particular patterns in the sky, and they do not fade as quickly as they used to. They also appear to contain shiny, rainbowlike particles within them.

This is because they are laced with chemicals that eventually shower down upon us all.

Like the thousands of other chemtrail spotters in the United States, Cheryl was mystified and concerned, but could only theorize as to what the chemicals were for. Mind control? Forced sterilization? Experimentation with new germ-warfare techniques? Then, when Japan was struck by an earthquake and tsunami in March 2011, so soon after that nation had rejected an American-led trade pact, she realized what was going on: the chemtrails are depositing highly conductive particles that allow for better long-distance functioning of the HAARP weather-controlling station in Alaska, run by the Defense Advanced Research Projects Agency (DARPA). As everyone knows, HAARP, the High Frequency Active Auroral Research Program, isn't just for researching the ionosphere, but for broadcasting signals that can change the weather, create earthquakes, and manufacture consent for the coming world government.

The host of the program calls her analysis "illuminating" and says he'll be having a chemtrail expert on very soon to help "tie together all these loose ends, and more."

Cheryl's theory of how it all fits together isn't unique. There are dozens of websites and YouTube videos making similar connections between the weather, military, economy, HAARP, natural disasters, and jet emissions. And they make up just a tiny fraction of the so-called conspiracy theories gaining traction online and in other media, connecting a myriad of loose ends, from 9/11 and Barack Obama's birthplace to the Bilderberg Group and immunizations.

They matter less for the solutions they come up with or the accusations they make than for the underlying need driving them all: to make sense of the world in the present tense. When there is no linear time, how is a person supposed to figure out what's going on? There's no story, no narrative to explain why things are the way they are. Previously distinct causes and effects collapse into one another. There's no time between doing something and seeing the result;

instead, the results begin accumulating and influencing us before we've even completed an action. And there's so much information coming in at once, from so many different sources, that there's simply no way to trace the plot over time. Without the possibility of a throughline we're left to make sense of things the way a character comes to great recognitions on a postnarrative TV show like *Lost* or *The Wire*: by making connections.

While we may blame the Internet for the ease with which conspiracy theories proliferate, the net is really much more culpable for the way it connects everything to almost everything else. The hypertext link, as we used to call it, allows any fact or idea to become intimately connected with any other. New content online no longer requires new stories or information, just new ways of linking things to other things. Or as the social networks might put it to you, "Jane is now friends with Tom." The connection has been made; the picture is getting more complete.

It's as if we are slowly connecting everyone to everyone else and everything else. Of course, once everyone is connected to everyone and everything else, nothing matters anymore. If everyone in the world is your Facebook friend, then why have any Facebook friends at all? We're back where we started. The ultimate complexity is just another entropy. Or as Cheryl put it, "Everything is everything."

The ease with which we can now draw lines of connectivity between people and things is matched only by our need to find patterns in a world with no enduring story lines. Without time, we can't understand things in terms of where they came from or where they are going to. We can't relate to things as having purpose or intention, beginnings or endings. We no longer have career paths, but connections and org charts. We don't have an economy of investments over time, but an economy of current relationships. We don't relate to the logic of sequential PowerPoint slides anymore, but to the zoomable canvas of Prezi, a presentation utility in which a single complicated picture is slowly revealed as the sum of many connected parts. We don't have a history of the world but a map of the world. A data visualization. A story takes time to tell; a picture exists in the static moment.

We can't create context in time, so we create it through links. This is connected to that. This reminds us of that. This reflects that. The entire universe begins to look holographic, where each piece somehow reflects the whole.

It's a sensibility we find reinforced by systems theory and chaos math. Fractals (those computer-rendered topologies that were to early cyberculture what paisley was to the 1960s) help us make sense of rough, natural phenomena, everything from clouds and waves to rocks and forests. Unlike traditional, Euclidean mathematics, which has tended to smooth out complexity, reducing it down to oversimplified lines and curves, fractal geometry celebrates the way real objects aren't really one, two, or three dimensions, but ambiguously in between.

Fractals are really just recursive equations—iterations upon iterations of numbers. But when they are rendered by computers, they churn out beautiful, complex patterns. They can look like a coral reef or a fern or a weather system. What makes fractals so interesting is that they are self-similar. If you zoom in on a shape in the pattern and look at the image at a much higher scale, you find that very same shape reappearing in the details on this new level. Zoom in again and the patterns emerge again.

On the one hand, this makes fractals terrifically orienting: as above, so below. Nature is patterned, which is part of what makes a walk in the woods feel reassuring. The shapes of the branches are reflected in the veins of the leaves and the patterns of the paths between the trunks. The repeating patterns in fractals also seem to convey a logic or at least a pattern underlying the chaos. On the other hand, once you zoom in to a fractal, you have no way of knowing which level you are on. The details at one level of magnification may be the same as on any other. Once you dive in a few levels, you are forever lost. Like a dream within a dream within a dream (as in the movie *Inception*), figuring out which level you are on can be a challenge, or even futile.

Meanwhile, people are busy using fractals to explain any system that has defied other, more reductionist approaches. Since they were successfully applied by IBM's Benoit Mandlebrot to the problem of seemingly random, intermittent interference on phone lines, fractals have been used to identify underlying patterns in weather systems, computer files, and bacteria cultures. Sometimes fractal enthusiasts go a bit too far, however, using these nonlinear equations to mine for patterns in systems where none exist. Applied to the stock market or to consumer behavior, fractals may tell less about those systems than about the people searching for patterns within them.

There is a dual nature to fractals: They orient us while at the same time challenging our sense of scale and appropriateness. They offer us access to the underlying patterns of complex systems while at the same time tempting us to look for patterns where none exist. This makes them a terrific icon for the sort of pattern recognition associated with present shock—a syndrome we'll call *fractalnoia*. Like the robots on *Mystery Science Theater 3000*, we engage by relating one thing to another, even when the relationship is forced or imagined. The tsunami makes sense once it is connected to chemtrails, which make sense when they are connected to HAARP.

It's not just conspiracy theorists drawing fractalnoid connections between things. In a world without time, any and all sense making must occur on the fly. Simultaneity often seems like all we have. That's why anyone contending with present shock will have a propensity to make connections between things happening in the same moment—as if there had to be an underlying logic. On the business-news channels, video footage of current events or presidential press conferences plays above the digital ticker tape of real-time stock quotes. We can't help but look to see how the president's words are influencing the Dow Jones average, as if sentiment on the trading floor really was reacting in live response to the news. Or maybe it is?

In an even more pronounced version of market fractalnoia, online business-news services such as the *Wall Street Journal*'s website or CBS MarketWatch rush to report on and *justify* stock market fluctuations. They strain to connect an upbeat report from the European Central Bank to the morning's 50-point rise—as if they know there is a connection between the two potentially unrelated events. By the time the story is posted to the Web, stocks are actually lower, and the agencies are hard at work searching for a housing report or consumer index that may explain the new trend, making the news services appear to be chasing their own tails.

This doesn't mean pattern recognition is futile. It only shows how easy it is to draw connections where there are none, or where the linkage is tenuous at best. Even Marshall McLuhan realized that a world characterized by electronic media would be fraught with chaos and best navigated through pattern recognition. This is not limited to the way we watch media but is experienced in the way we watch and make choices in areas such as business, society, and war.

Rapid churn on the business landscape has become the new status quo, as giants like Kodak fall and upstarts like Facebook become more valuable than oil companies. What will be the next Zynga or Groupon? Surely it won't be another social-gaming or online-coupon company—but what would be their equivalent in the current moment? How do we connect the dots? On the geopolitical stage, war no longer occurs on battlefields or in relationship to some diplomatic narrative, but as a series of crises, terrorist attacks, or otherwise disproportionate warfare. Where and when will the next attack take place?

The trick is to see the shapes of the patterns rather than the content within them—the medium more than the message. As I have come to understand the cultural swirl, media events tend to matter less for whatever they are purportedly about than for the space they fill. Charlie Sheen did not rise to Twitter popularity merely by being fired from his sitcom and posting outlandish things; he was filling an existential vacuum created in the wake of the Arab Spring story immediately preceding him. In effect our highly mediated culture creates a standing wave; the next suitable celebrity or story that comes along just happens to fill it. Or, as I explained in my book *Media Virus*, the spread of a particular virus depends no more on the code within the virus than it does on the immune response of the culture at large.

However counterintuitive this may seem: sometimes the best way to see where things are going is to take our eyes *off* the ball. We don't identify the next great investment opportunity by chasing the last one, but by figuring out what sorts of needs that last one was fulfilling, and the one before that. We don't predict the next suicide bombing by reading the justifications for civil unrest in a foreign-policy journal but by inferring the connections between nodes in the terror network. Like a surfer riding the tide, we learn to look less at the water than at the waves.

In doing so, and as we'll see in this chapter, we quickly realize that pattern recognition is a shared activity. Just as the three-dimensional world is best perceived through two eyes, a complex map of connections is better understood from more than one perspective at a time. Reinforcing self-similarity on every level, a network of people is better at mapping connections than a lone individual. As author and social critic Steven Johnson would remind us, ideas don't generally emerge from individuals, but from groups, or what he calls "liquid networks."[1] The coffeehouses of eighteenth-century London spawned the ideas that fueled the Enlightenment, and the Apple computer was less a product of one mind than the collective imagination of the Homebrew Computer Club to which both Steve Jobs and Steve Wozniak belonged.

The notion of a lone individual churning out ideas in isolated contemplation like Rodin's Thinker may not be completely untrue, but it has certainly been deemphasized in today's culture of networked thinking. As we become increasingly linked to others and dependent on making connections in order to make sense, these new understandings of how ideas emerge are both helpful and reassuring. For example, researcher Kevin Dunbar recorded hundreds of hours of video of scientists working in labs. When he analyzed the footage, he saw that the vast majority of breakthroughs did not occur when the scientists were alone staring into their microscopes or poring over data, but rather when they were engaged with one another at weekly lab meetings or over lunch.[2]

This sort of observation provides a bit of comfort to those of us slow to warm to the fact that thinking is no longer a personal activity, but a collective one. Nothing's personal—except maybe the devices through which we connect with the network. New ideas seem to emerge from a dozen places at once, a mysterious zeitgeist synchronicity until we realize that they are all aspects of the same idea, emerging from a single network of minds. Likewise, each human brain is itself a network of neurons, sharing tasks and functioning holographically and nonlocally. Thoughts don't belong to any one cell any more than ideas belong to any one brain in the greater network of a connected human culture.

The anxiety of influence gives way to the acceptance of intimacy and shared credit. Many young people I encounter are already more than comfortable losing their privacy to social networks, preferring to see it as preparation for an even less private, almost telepathic future in which people know one another's thoughts, anyway. In a networked ideascape, the ownership of an idea becomes as quaint and indefensible a notion as copyright or patents. Since ideas are built on the logic of others, there is no way to trace their independent origins. It's all just access to the shared consciousness. Everything is everything. Acceptance of this premise feels communist or utopian; resistance feels like paranoia.

Regardless of whether this culture of connectivity will bring us to greater levels of innovation and prosperity, this culture is certainly upon us. Learning how to recognize and exploit patterns without falling into full-fledged fractalnoia will soon be a required survival skill for individuals, businesses, and even nations.

THE FEEDBACK LOOP: PARSING SCREECH

"That Tweet could be the beginning of the end," the CEO told me, slowly closing his office door as if to mask the urgency of the situation from workers in cubicles outside. In less than 140 characters, a well-followed Tweeter was able to foist an attack on a corporation as disproportional and devastating as crashing a hijacked plane into an office tower.

The details of the scandal are unimportant. It turned out the Twitter-amplified attack on the company was actually initiated by the improper use of hashtags by the company's own hired online publicity firm, anyway. But the speed with which error became a virus and then spread into a public relations nightmare covered by mainstream media was shocking. At least it was shocking to a CEO accustomed to the nice, slow pace of change of traditional media. Back then, a company would be asked for a comment, or a CEO could pick up the phone and talk to a few editors before a story went out. A smart PR firm could nip it in the bud.

Now there is no bud. Just pollen. Everywhere. It's not merely a shift in the pace of media—as when things shifted from delivered print newspapers to broadcast TV—but in the direction of transmission. And even then, it's more complex and chaotic then we may first suspect. To an imperiled politician or CEO, the entire world seems to have become the enemy. Everything is everything.

For many of us, this goes against intuition. We like to think having more connections makes us more resilient. Isn't more friends a good thing? Yes, there is strength in networks—particularly grass-roots networks that grow naturally over time and enjoy many levels of mutual support. But all this connectedness can also make us *less* resilient.

In a globally connected economy, there is no such thing as an isolated crash. It used to be that the fall of one market meant another was going up. Now, because they are all connected, markets cannot fall alone. A collapse in one small European nation also takes down the many overseas banks that have leveraged its debt. There's nowhere to invest that's insulated from any other market's problems. Likewise, thanks to the interconnectedness of our food supply and transportation networks, the outbreak of a disease on the poultry farms of China necessarily threatens to become a global pandemic. A connected world is like a table covered with loaded mousetraps. If one trap snaps, the rest of the table will follow in rapid, catastrophic succession. Like a fight between siblings in the back of the car on a family trip, it doesn't matter who started it. Everybody is in it, now.

Along with most technology hopefuls of the twentieth century, I was one of the many pushing for more connectivity and openness as the millennium approached. It seemed the only answer for our collapsing, top-down society was for everyone and everything to network together and communicate better and more honestly. Instead of emulating a monarchy or a factory, our society could emulate a coral reef—where each organism and colony experiences itself as part of a greater entity. Some called it Gaia, others called it evolution, others called it the free market, and still others called it systems theory. Whatever the metaphor, a connected world would respond more rapidly and empathically to crises in remote regions, it would become more aware of threats to its well-being, and may actually become more cooperative as a whole. It seemed possible that a networked human society of this type may even have some thoughts or purpose of its own.

But the Earth reached its maximum state of openness, at least so far, on September 10, 2001. The next day, we learned how all this openness can be turned on its head, lending the explosive impact of jet liners and the population density of modern skyscrapers to a few guys armed with box cutters. That's when the dream of a connected world revealed itself rather as the nightmare of a world vulnerable to network effects. Whether it's a congressman losing his job over an errant Tweet, a State Department humiliated by the release of its cache of cables on WikiLeaks, or a corporation whose progressive consumers just learned their phones were assembled in Chinese sweatshops, connectedness has turned every glitch into a potentially mortal blow, every interactor into a potential slayer.

It may seem glib to equate a terrorist attack with a public relations snafu, but from the perspective of the institutions now under seemingly perpetual assault, it is the same challenge: how can they better respond to the crises emerging seemingly from anywhere and everywhere, all the time? How can they predict and avoid these crises in the first place? When everything is connected to everything else, where does this process even begin? Any action from anywhere may be the game changer—the butterfly flapping its wings in Brazil that leads to a hurricane in China. And all of it happening at the same time, because cause and effect seem to have merged into the same moment.

Welcome to the world of feedback and iteration, the cyclical processes on which fractals are based. In a normal situation, feedback is just the response you get from doing something. It's the echo that comes back after you shout into a canyon, the quizzical looks on the faces of students after you tell them a new idea, or even the size of the harvest you get after trying a new fertilizer. You take an action and you get feedback. You then use this feedback to figure out how to adjust for the next time. You shoot an arrow; it lands to the left of the bull's eye; you adjust by aiming farther to the right; and so on. Each time you adjust is another iteration of the feedback loop between you and the target.

For feedback to be useful there must be some interval between the thing you've done and the result you've created. You need time to see what happened and then adjust. In a presentist world, that feedback loop gets really tight. Feedback happens so fast that it becomes difficult even to gauge what's happening. You know that feeling when you're holding a microphone and the speakers suddenly *screech*, and you don't know which way to move to make it stop? That's actually feedback you're dealing with and trying to control. Normally the microphone simply hears the sound of your voice and passes it on to the amplified speaker. When you get too close to one of those speakers, however, the microphone ends up listening to its own noise. Then that noise goes back through the amplifier and speaker, at which point the microphone hears it again, and sends it back on through—again, and again, and again. Each iteration amplifies the sound more and more, thousands of times, until you hear the combined, chaotic screech of an infinite and instantaneous feedback loop. Only we don't hear a cyclical loop; we just hear the high-pitched whine. That's the way the world sounds right now to most governments and businesses. Everything becomes everything.

Fractals are really just a way of making sense of that screech. Deep inside that screech is the equivalent of one of those cyclical, seemingly repetitive Philip Glass orchestral compositions. We just don't have the faculties to hear it. Computers, on the other hand, work fast enough that they have time to parse and iterate the equation. Like a kid drawing seemingly random circles with a Spirograph, computers track the subtle differences between each feedback loop as it comes around, until they have rendered the utterly beautiful tapestries that evoke coral reefs, forest floors, or sand dunes, which are themselves the products of cyclical iterations in the natural world.

The fractal is the beautiful, reassuring face of this otherwise terrifying beast of instantaneous feedback. It allows us to see the patterns underlying the seeming chaos, the cycles within the screechy collapsed feedback of our everything-all-at-once world.

Feedback used to be slow. A company might upgrade a product and put it out into the market, and wait a full season to see how it did. Feedback came in the form of inventory reports and returns. The way a product sold was the feedback a company needed to plan for the next season. In traditional politics, feedback came every few years through the voting booth. It eventually tightened to polls conducted weekly or even daily, but at least there was some control and sequence to the process: leak a policy, poll the public, then either announce it or not.

When feedback comes instantaneously and from all sides at once, it's hard to know how people are reacting to what we are doing—or what we're doing that they're even reacting to. Social media lets people feed back their responses immediately and to one another instead of just back to the business or politician concerned. Then other people respond as much to those messages as they do to the product or policy. They are feeding back to one another. In a landscape with instantaneous lateral feedback, marginal box office on the opening night of a movie—even if it has nothing to do with the movie—ends up being Tweeted to the next day's potential viewers. The negativity iterates, and the movie fails, even if it got good reviews from the traditional channels of feedback.

Businesses attempt to adjust in real time to the comments about their products emerging on bulletin boards, update feeds, or anywhere else, and often don't even know whether the feedback is coming from someone who has seen or touched the actual product. Brand managers know they're in a world where companies are expected to be present in all these venues, responding and adjusting, clarifying and placating—but it's hard to do when everybody is talking at once, feeding back into one another and the product and the competition and the shareholders. Thanks to feedback and iteration, any single Tweet can mushroom into a cacophony. Ideation, corporate culture, development, production, branding, consumer research, and sales all become part of an iterative, circular equation where causes and effects can no longer be parsed. When feedback comes through the cycle in that uncontrollable way, it's like putting that microphone next to the speaker amplifying its own signal. All you get is screech. You don't know where to stand to make it stop. The speakers are everywhere. Everything is everything.

Traditional corporate communications don't function in such an environment. Remember, corporations themselves were born to counteract the peer-to-peer, every-which-way connectivity of the bazaar. As we saw earlier, the corporate brand emerged as a stand-in for the community member who may have once provided the goods that are now being shipped in from some faraway place. The company doesn't want us dwelling on the factories where our cookies are being made, especially when they could just as easily have been baked by a friend in town without the need for long-distance shipping or unhealthy preservatives. So instead of telling us the true story of the factory, the corporation tells us the story of how elves baked the cookies in a hollow tree. This is what is known as brand mythology; it was not developed to enhance the truth, but to replace it.

Now this all worked well enough in an age of print and broadcast media. Those media were themselves developed largely to promote this sequential, back-and-forth style of communication between corporations and their consumers. In fact, the whole notion of balanced reporting by newspapers—itself a mythological construct if ever there was one—emerged in response to the advertising needs of corporations, who did not want their ads appearing next to editorial copy that might alienate any potential customers. Television, similarly, served as little more than electronic wallpaper for advertising. Difficult ideas were left for the movies, PBS, and, later, HBO.

Until quite recently, companies enjoyed exclusive dominion

over the vast majority of messaging that reached people. This allowed them to craft the stories they wanted to, control their dissemination through the media, and then respond to the orderly sales feedback or controlled sentiment studies. The rise of interactive technologies, which finally gave consumers a way to feed back directly to their screens in real time, encouraged many companies to begin thinking less in terms of story than in those of "conversation."

As Kevin Roberts, CEO of the once colossal Saatchi & Saatchi, explained to me, "The consumer is now in total control." He waxed sentimental as he described the new empowerment. "I mean she can go home, she's going to decide when she buys, what she buys, where she buys, how she buys. . . . Oh, boy, they get it you know, they're so empowered at every age. They are not cynical; they are completely empowered; they're autonomous. All the fear is gone and all the control is passed over to the consumer. It's a good thing."[3]

Many companies tried to take part in this conversation, under the impression that consumers really wanted to speak with them. What they failed to recognize is that consumers don't want to speak with companies through social media; we want to speak with one another. We don't even think of ourselves as consumers anymore, but as *people*. The great peer-to-peer conversation of the medieval bazaar, which was effectively shut down by the rise of corporate communications, is back. One advantage (or side effect, depending on your perspective) of being always-on is that we are never limited to just one role. Thanks to smart phones, we are available to our clients while vacationing with our family at the beach. We bid on eBay items from work and cruise for prospective mates from the safety of our own bedrooms. On social networks, we are at once consumer, producer, citizen, parent, and lover.

Moreover, people are not engaging with one another over Twitter and Facebook about elves who make cookies or bears who soften laundry. They are not telling stories in 140 characters or less, but sharing facts. Updates concern things that matter in the present tense: What's really in this cookie? Did you hear what I found out about the factory they're made in? Do the chemicals in this fabric softener get absorbed through a baby's skin? When people are concerned with questions like these, brand mythologies cease to have any relevance—except when they serve as ironic counterpoint to the facts on the ground. Instead, people compete to provide one another with valuable information or informed opinions as a way of gaining popularity, or *social currency*, in their networks. They feedback not just to companies or governments, but to one another.

So, as we have seen, narrative has collapsed, branding has become irrelevant, consumers see themselves as people, and everyone is engaged in constant, real-time, peer-to-peer, nonfiction communication. All the while, companies are busy trying to maintain linear, call-and-response conversations about brand mythologies with consumers. In such an environment, it's no wonder a tiny gaffe can overtake many years and dollars of strategy. Amplified by feedback and iterated ad infinitum, a tiny pinprick of truth can pop a story that took decades to inflate.

And corporations are not alone. Political campaigns, governments, foundations, and religious institutions—which have all adopted the communications and operational styles of corporations—face the very same vulnerabilities in an everything-is-everything cacophony.

Those who think they are hip to the shifting ground rules invite their constituencies inside, but only to participate in more myth creation. For example, Hillary Clinton's failed presidential run was characterized chiefly by faux efforts to incorporate feedback into her campaign. She began with a "listening tour" through the nation—during which she sat and nodded sympathetically while regular people told her what was wrong with their lives. This had nothing to do with creating policy and everything to do with creating the appearance of responding to feedback. As a result of all she heard, she "decided" to run for president. Then, pounding on the feedback theme still further, she invited her constituency to choose her campaign theme song by "voting" for one of the selections on her website. Her campaign ventured futilely into the era of feedback by aping the democratic integrity of *American Idol*. (She later announced the winning song in a misguided and much-ridiculed parody of *The Sopranos*—as if to demonstrate facility with the everything-is-everything resonances of a self-reflexive mediaspace. All she succeeded in doing was to equate the Clinton dynasty with that of a fictional mob family. Even if she wasn't fractalnoid, her audience was busy making the connections.)

Corporations are attempting to enter the feedback loop as well—or at least trying to create limited opportunities for controlled feedback to occur. Crowdsourcing is really just a corporation's way of trying to focus the otherwise random feedback from consumers onto a particular task. Unfortunately, they're doing so without fully considering the liabilities. General Motors, for example, invited consumers to make commercials online for one of its SUVs. The company developed a very sophisticated set of Web utilities through which users could select footage, edit, add music, write title cards, and create special effects. It was a supreme statement of trust and consumer empowerment. But instead of making commercials for the vehicle, more creative participants produced videos criticizing the gas-guzzling Chevy Tahoe and the way they saw GM equating machismo and patriotism with wasteful energy consumption.[4] Web visitors quickly voted these to the top of the favorites list, where they garnered the attention of television news shows. Indeed, the campaign went more viral than GM could have hoped. *Screech.*

To GM, who eventually pulled down the website, this must have seemed like an assault, some sort of activist prank or media terrorism. Why was everyone attacking GM, when all the company had done was offer people the chance to make some videos? Wasn't this how crowdsourcing is supposed to work? For their part, the people who made the videos were simply using the tools GM provided and beginning the conversation that the company said it wanted to have. Again, the limits of openness had been reached, feedback iterated spontaneously and instantaneously into screech, and another company rethought whether it wanted to have a social media strategy at all.[5]

Of course, in a landscape where everyone is connecting and feeding back to everything and everyone else, there is no such thing as not having a social media strategy. People inside and outside every organization—even clandestine ones—are still engaging with friends, colleagues, competitors, and strangers online. They are happy or unhappy with their jobs, purchases, representatives, schools, banks, and systems of government. Except for the few who are paid to do otherwise (professional online shills), most are telling the truth—or at least revealing it through silence, subtle cues, or their data trails. Things stay open, anyway.

An official social media strategy, executed by a professional PR firm, may help an organization deal with explicit complaints about its products. There are filtering services that can scour the net for comments made to almost any feed, giving clients the chance to answer complaints or accusations appropriately within minutes. That a company in a presentist universe must be always-on and ready to respond to critique instantaneously should go without saying. As long as these Tweets, updates, and posts are limited to a few thousand a day, this remains a manageable proposition.

But this approach is still a carryover from the days of broadcast media and easy, top-down control of communications. Back in the era of television and other electronic communications technologies, a "global media" meant satellite television capable of broadcasting video of the Olympics across the globe. This was the electronically mediated world Marshall McLuhan described as the "global village"; he was satirizing the hippy values so many thought would emerge from a world brought together by their TV sets, and in his way warning us about the impact of globalism, global markets, and global superpowers on our lives and cultures. With the rise of digital media, however, we see the possibility for a reversal of this trend. Unlike the broadcast networks of the electronic age, digital networks are biased toward peer-to-peer exchange and communication. Instead of big institutions responding to (and, in some cases, mitigating) the feedback of a world of individuals, those individuals are feeding back to one another. The institutions aren't even in the conversation.

Instead of simply responding to feedback from consumers or constituents, institutions contending with a peer-to-peer mediaspace must stop "messaging" and instead just give people the facts and fuel they need to engage with one another in a manner that helps everyone. This means thinking of one's customers, employees, shareholders, and competitors a bit more holistically. Social media is still social, and users move between their various identities interchangeably. The off-duty computer repairperson may be telling people more about the relative durability of laptops than the technology-review website does. Prospective shareholders engage with customers, customers engage with employees, employees engage with competitors, and competitors engage with suppliers and partners. And not all their communications will have an organization's name in the subject field. Anything and everything that everyone is doing right now may matter. There is no damage control or crisis management. There's just what's happening now.

The only real choice is to give up and make that "now" as truly beneficial to an organization's goals as possible. As strange or even naive as it may sound, this means abandoning communications as some separate task, and instead just *doing* all the right things that you want talked about. The sheer volume, constancy, and complexity of communications are too hard to consciously manage anymore. They must be regarded as the expressions of a living culture whose growth and fertility are inextricably linked to one's own.

This is easier said than done. I regularly receive calls for help from companies and organizations looking to become more transparent. The trouble is, many of them don't really do anything they can reveal to the public. An American television manufacturer that wanted to "get more social" with its communications strategy didn't realize this would be impossible now that it no longer designs or makes its television sets. (It outsourced both functions to others.) Who is left to represent the company through its online interactions? Likewise, a politician wanted to become identified through social media as more of a "hometown hero"—even though he had lived in the district where he was running for only a few weeks!

In 2004 Congress authorized the funding of an international cable news channel called Alhurra ("the free one"), headquartered in Virginia but broadcast in Arabic throughout the Arab-speaking world. Launched in the wake of 9/11 at a cost of over $100 million per year, the US government's channel was supposed to help forge better relations with viewers in these potentially hostile parts of the world. In spite of this massive effort for Alhurra to become the trusted, balanced news channel for Arabic-speaking countries, the target audiences immediately realized the propagandistic purpose of the channel and the true culture from which it emanated. Meanwhile, the English-speaking version of Arab station Al Jazeera became an unintentional hit in the United States during the Arab Spring and Occupy Wall Street—simply because it broadcast its continuous coverage live online and showed journalistic competence in the field. (Hundreds of thousands of Americans still watch Al Jazeera, and talk about it, and interact with its broadcasters—and will likely continue to do so until a major American cable news channel figures out how to get around cable carrier contracts limiting net streaming.)

In a nonfiction, social media space, only reality counts, because only reality is what is happening in the moment. A company or organization's best available choice is to walk the walk. This means becoming truly competent. If a company has the best and most inspired employees, for example, then that is the place people will turn to when they are looking for advice, a new product, or a job. It is the locus of a culture. Everything begins to connect. And when it does, the org chart begins to matter less than the fractal of fluid associations.

The employees and customers of the computer-game-engine company Valve are particularly suited to experiment with these principles. The privately owned company's flagship product, the video game platform Steam, distributes and manages over 1,800 games to a worldwide community of more than 40 million players. They approach human resources with same playfulness as they do their products, enticing website visitors to consider applying for a job at the company: "Imagine working with super smart, super talented colleagues in a free-wheeling, innovative environment—no bosses, no middle management, no bureaucracy. Just highly motivated peers coming together to make cool stuff. It's amazing what creative people can come up with when there's nobody there telling them what to do."[6]

This is both a hiring tactic and good publicity. Customers want to believe the games they are playing emerge from a creative, playful group of innovators. But for such a strategy to work in the era of peer-to-peer communications, the perception must also be true. Luckily for Valve, new-media theorist Cory Doctorow got ahold of the company's employee manual and published excerpts on the tech-culture blog BoingBoing. In Doctorow's words, "Valve's employee manual may just be the single best workplace manifesto I've ever read. Seriously: it describes a utopian Shangri-La of a workplace that makes me wish—for the first time in my life—that I had a 'real' job."[7]

Excerpts from the manual include an org chart structured more like a feedback loop than a hierarchy, and seemingly unbelievable invitations for employees to choose their own adventures: "Since

Valve is flat, people don't join projects because they're told to. In-stead, you'll decide what to work on after asking yourself the right questions. . . . Employees vote on projects with their feet (or desk wheels). Strong projects are ones in which people can see demon-strated value; they staff up easily. This means there are any number of internal recruiting efforts constantly under way. If you're working here, that means you're good at your job. People are going to want you to work with them on their projects, and they'll try hard to get you to do so. But the decision is going to be up to you."[8] Can a com-pany really work like that? A gaming company certainly can, espe-cially when, like Valve, it is privately owned and doesn't have shareholders to worry about. But what makes this utopian work-place approach easiest to accept is the knowledge that one's employ-ees embody and exude the culture to which they're supposedly dedicated. They are not responding to game culture, but rather cre-ating it.

The fractal is less threatening when its shapes are coming from the inside out. Instead of futilely trying to recognize and keep up with the patterns within the screech—which usually only leads to paranoia—the best organizations create the patterns and then enjoy the ripples. Think of Apple or Google as innovators; of Patagonia or Herman Miller as representing cultures; of the Electronic Frontier Foundation or Amnesty International as advocating for constituen-cies; of Lady Gaga or Christopher Nolan as generating pop culture memes. They generate the shapes we begin to see everywhere.

In a social world, having people who are capable of actually generating patterns is as important for a church or government agency as it is for a corporation or tech start-up. They do some-thing neat, then friends tell friends, and so on. If an organization al-ready has some great people, all it needs to do is open up and let them engage with the other great people around the world who care. Yes, it may mean being a little less secretive about one's latest innovations—but correspondingly more confident that one's greatest innovations still rest ahead.

The examples go on and on, and surely you know many more yourself. Organizations that focus on controlling everyone's response to them come off like neurotic, paranoid individuals; there's just too much happening at once to second-guess everyone. Those who dedicate their time and energy to their stakeholders end up becoming indistinguishable from the very cultures they serve—and just as fertile, interconnected, complex, and alive.

The fractal acts like a truth serum: the only one who never has to worry about being caught is the one who never lied to begin with.

MANAGING CHAOS: BIRDS, BEES, AND ECONOMIES

The hyperconnected fractal reality is just plain incompatible with the way most institutions operate—especially governments, whose exercises in statecraft are based on a cycle of feedback that seems more tuned to messages sent by carrier pigeon than the Internet. Keeping an eye on a message as it spins round through today's infinitely fast feedback loops only makes them dizzy.

The problem is we are focused on the object instead of the motion, or, as McLuhan would put it, the figure instead of the ground—Charlie Sheen instead of the standing wave he entered. We need to unfocus our eyes a bit to take in the shape of what's happening. Stepping back and looking at the picture is fraught with peril, too, however—as we are so tempted to draw premature connections between things. We think of this sort of pattern recognition and lateral thinking as less logical and more intuitive, more "stoned" than "straight." This is because most of us, at least in the West, are too inexperienced beholding things this way to do so with any rigor or discipline. Once we start seeing how things are connected, we don't know how to stop. We may as well be on an acid trip, beholding the web of life or the interplay of matter and energy for the very first time.

This is why, when confronted with the emerging complexity of the twentieth century, the initial response of government and business alike was to simplify. They put their hands over their ears so as not to hear the rising screech and looked to their models, maps, and plans instead. Like generals in the safety of a situation room using toy tanks on a miniature battlefield to re-create a noisy and violent war, government and corporate leaders strategized in isolation from the cacophony of feedback. They sought to engage with their challenges from far above and to make moves as if the action were occurring on a game board. Complexity was reduced to simple, strategic play.

This is what "gaming" a system really means, and what kept Cold War powers busy building computers for the better part of the twentieth century. After all, the invention and use of the atomic bomb had taught us that we humans might be more connected than we had previously realized. In an early form of fractalnoia, we came to realize that a tiny misstep in Cuba could lead to a global thermonuclear war from which few humans survive.

Gaming gave leaders a way to choose which feedback to hear and which to ignore. Even if there were millions of possible actors, actions, and connections, there were only two real superpowers—the Soviet Union and the United States. Military leaders figured that game theory, based on the mathematics of poker, should be able to model this activity and give us simple enough rules for engagement. And so the RAND Corporation was hired to conduct experiments (like the Prisoner's Dilemma, which we looked at earlier), determine probable outcomes, and then program computers to respond appropriately in any number of individual circumstances.

Led by the as yet undiagnosed paranoid schizophrenic John Nash (the mathematician portrayed in the movie *A Beautiful Mind*), they adopted a principle called MAD, or mutually assured destruction, which held that if the use of any nuclear device could effectively guarantee the complete and utter annihilation of both sides in the conflict, then neither side would opt to use them. While this didn't stop the superpowers from fighting smaller proxy wars around the world, it did serve as a deterrent to direct conflict.

Encouraged by this success, Nash applied his game theory to all forms of human interaction. He won a Nobel Prize for showing that a system driven by suspicion and self-interest could reach a state of equilibrium in which everyone's needs were met. "It is understood not to be a cooperative ideal," he later admitted, but—at least at the time—neither he nor RAND thought human beings to be cooperative creatures. In fact, if the people in Nash's equations attempted to cooperate, the results became much more dangerous, messy, and unpredictable. Altruism was simply too blurry. Good planning required predictable behaviors, and the assumption of short-term self-interest certainly makes things easy to see coming.

A few decades of game theory and analysis since then have revealed the obvious flaws in Nash's and RAND's thinking. As Hungarian mathematician and logician László Mérő explains it in his rethink of game theory, *Moral Calculations,*[9] the competitive assumptions in game theory have not been proved by consistent results in real-world examples. In study after study, people, animals, and even bacteria are just as likely to cooperate as they are to compete. The reason real human behavior differs from that of the theoretically self-interested prisoners is that the latter are prisoners to begin with. An incarcerated person is the most literal example of one living within a closed environment. These are individuals without access to information and incapable of exercising basic freedoms. All feedback and iteration are removed, other than that between the prisoner and his keepers. With the benefit of several

hundred prisoner dilemma studies to mine for data and differences, Mérő found that the more the "prisoners" knew about their
circumstances and those of their fellow prisoners, the less selfishly
they behaved. Communication between prisoners invariably yielded
more cooperation. Isolation bred paranoia, as did more opacity
about how the rules worked. Communication, on the other hand,
generates lateral feedback loops and encourages a more extended
time horizon.

Mérő's research into humans and other biological systems dem-
onstrates that species do conduct game-theory-like calculations about
life-or-death decisions, but that they are not so selfishly decided as
in a poker game. Rather, in most cases—with enough lateral feedback at their disposal—creatures employ "mixed strategies" to
make decisions such as "fight or flight." Further, these decisions uti-
lize unconsciously developed (or perhaps instinctual) probability
matrices to maximize the survival of the species and the greater
ecosystem. In just one such cooperative act, the competition over a
piece of food, many species engage in threatening dances instead of
actual combat. By cooperatively negotiating in this fashion—and
using battle gestures, previous experiences, and instinct (species'
memory) to calculate the odds of winning or losing the battle—
both individuals live on to see another day.

In contrast, game-theory tests like the Prisoner's Dilemma set
up competitions where decisions are forced with a lack of information. They are characterized by noncommunication. There is no
space for negotiation or transparency of decision making, nor any
participation in extending the range of possible outcomes. The prisoners are operating in the least likely circumstances to engender cooperative actions.

But the zero-sum logic of game theory can still work, as long as
it is actualized in a culture characterized by closedness. This is why
corporations functioning in this fashion gain more power over la-

borers who are competing rather than unionizing; it's why real estate agents can jack up prices more easily in a market where "comps"
are not available; and it's why a larger government can push around
developing nations more easily when its cables aren't being posted on
the Internet by WikiLeaks. Less networking and transparency keeps
everyone acting more selfishly, individualistically, and predictably.

In the controlled information landscape, these strategies worked
pretty well for a long time. A closed, top-down broadcast media
gave marketers and public relations specialists a nation of individuals with whom to communicate. *You, you're the one*—or so the commercials informed each of us. Suburban communities such as Levittown
were designed in consultation with Roosevelt administration psy-
chologists, to make sure they kept us focused inward.[10] Separate
plots and zoned neighborhoods reified the nuclear family at the ex-
pense of wider, lateral relationships between families. Decades of social control—from corporate advertising to manufacturing public
consent for war—were exercised through simple one-to-many campaigns that discouraged feedback from them and between them.
As long as people didn't engage with one another and were instead
kept happily competing with one another, their actions, votes, and
emotions remained fairly predictable. Screech could be kept to a
minimum.

But the Cold War gave rise to something else: a space race, and
the unintended consequence of the first photographs of planet Earth
taken from the heavens. Former Merry Prankster Stewart Brand had
been campaigning since 1966 for NASA to release a photo of Earth,
aware that such an image could change human beings' perception of
not only their place in the universe but also their relationship to one
another. Finally, in 1972, NASA released image AS17-148-22727,
birthing the notion of our planet as a "big blue marble." As writer
Archibald MacLeish described it, "To see the Earth as it truly is,
small and blue and beautiful in that eternal silence where it floats, is

to see ourselves as riders on the Earth together, brothers on that bright loveliness in the eternal cold—brothers who know now that they are truly brothers."[11]

Soon after that, the development of the Internet—also an outgrowth of the Cold War funding—concretized this sense of lateral, peer-to-peer relationships between people in a network. Hierarchies of command and control began losing ground to networks of feedback and iteration. A new way of modeling and gaming the activities of people would have to be found.

The idea of bringing feedback into the mix came from the mathematician Norbert Wiener, back in the 1940s, shortly after his experiences working for the military on navigation and antiaircraft weapons. He had realized that it's much harder to plan for every eventuality in advance than simply to change course as conditions change. As Wiener explained it to his peers, a boat may set a course for a destination due east, but then wind and tides push the boat toward the south. The navigator reads the feedback on the compass and corrects for the error by steering a bit north. Conditions change again; information returns in the form of feedback; the navigator measures the error and charts a new course; and so on. It's the same reason we have to look out the windshield of our car and make adjustments based on bumps in the road. It's the reason elevators don't try to measure the distance between floors in a building, but instead "feel" for indicators at each level. It's how a thermostat can turn on the heat when the temperature goes down, and then turn it off again when the desired temperature is reached. It feels.

Wiener understood that if machines were ever going to develop into anything like robots, they would need to be able to do more than follow their preprogrammed commands; they would have to be able to take in feedback from the world. This new understanding of command and control—*cybernetics*—meant allowing for real-time measurements and feedback instead of trying to plan for every emergent possibility. Wiener saw his new understandings of command

and control being used to make better artificial limbs for wounded veterans—fingers that could feed back data about the things they were touching or holding. He wanted to use feedback to make better machines in the service of people.

Social scientists, from psychologist Gregory Bateson to anthropologist Margaret Mead, saw in these theories of feedback a new model for human society. Instead of depending so absolutely on RAND's predictive gamesmanship and the assumption of selfishness, cybernetics could model situations where the reality of events on the ground strayed from the plan. Qualitative research, polling, and focus groups would serve as the feedback through which more adaptive styles of communication and—when necessary—persuasion could be implemented. During World War II and afterward, the society of Japan became the living laboratory for America's efforts in this regard—an exercise in "psychological warfare" Bateson would later regret and blame, at least in part, for the breakup of his marriage to Mead, who he was saddened to recall remained more committed to manipulative government agendas such as social control and militarism.[12]

Bateson wasn't really satisfied with the mathematics of simple feedback, anyway, and longed for a more comprehensive way to make sense of the world and all the interactions within it. He saw the individual, the society, and the natural ecology as parts of some bigger system—a supreme cybernetic system he called "Mind," which was beyond human control and to which human beings must in some sense surrender authority. We'll look at the religious implications of all this complexity and systems theory in the next chapter. What's important to this part of our story is that instead of applying cybernetics to mechanical systems in order to serve humans, Bateson was equating human society with a cybernetic system.

This model of humanity as some sort of complex system proved irresistible for a new generation of social scientists and economists alike, who had been looking for a way to apply nonlinear math and computing brawn to the problem of seemingly chaotic human activity—and particularly economics. The Cold War had grown from a purely military game to an ideological one. From the ivory towers of America's universities to the hearing rooms of the House Committee on Un-American Activities, capitalism was being defended and promoted as a way of life. During this period, the American intelligentsia as well as its funders—from the government to the Ford and Rockefeller foundations—were busily looking for scientific and mathematical proof of capitalism's supremacy. They wanted to prove that a free market could spontaneously deliver the order and fairness for which centrally planned communist bureaucracies could only strive in vain.

Adam Smith had already observed back in the 1700s how free markets did a great job of coordinating people's actions. They seemed to self-regulate even though no one was aware of anyone else's intentions. This "invisible hand" of the marketplace ensured that as individuals try to maximize their own gains in a free market, prices find the appropriate levels, supply meets demand, and products improve. But Smith only got so far as to argue that, through the market, competition channels individual ambitions toward socially beneficial results. The mechanisms through which that happened were still poorly understood and expressed.

In 1945 famed Austrian economist Friedrich Hayek came up with a new way of talking about the "invisible hand" of the marketplace. Instead of seeing some magical, inexplicable method through which markets solved its problems fairly, Hayek saw markets working through a more systemic process. In his view, pricing is established collectively by thousands or even millions of individual actors, each acting on his own little bit of knowledge. The marketplace serves to resolve all these little pieces of data in a process Hayek called "catallaxy"—a self-organizing system of voluntary cooperation. Feedback and iteration. Price mechanisms were not human inventions so much as the result of collective activity on a lower order. Humans were part of a bigger system that could achieve "spontaneous order."

Between chaos theory, cybernetic systems, and computing brawn, economists finally had the tools to approach the marketplace as a working catallaxy—as a product of nature as complex and stable as human life itself. Economists at the Santa Fe Institute, in particular, churned out chaotic models for the collective economic activity to which humans contributed unconsciously as members of this bigger fractal reality. Founded in 1984, the institute encourages the application of complex adaptive systems to work in the physical, biological, computational, and social sciences. Their research projects span subjects from "the cost of money" and "war-time sexual violence" to "biological systems" and "the dynamics of civilizations," applying the same sorts of computational models for nonlinear dynamical systems to all these different levels of reality.

The key to any of this working was what became known as emergence—the spontaneous achievement of order and intelligence through the interaction of a myriad of freely acting individuals. Birds do it, bees do it . . . free market economies do it. And now we have the fractals with which to catch them all in the act.

Scientists from across the spectrum leaped on the systems bandwagon, applying what began as a mathematical proof of market equilibrium to, well, pretty much everything. Linguist Steven Pinker saw in Hayek and systems theory a new justification for his advancement of evolutionary psychology and his computational theory of mind:

"Hayek was among the first to call attention to the emergence of large-scale order from individual choices. The phenomenon is ubiquitous, and not just in economic markets: What makes everyone suddenly drive SUVs, name their daughters Madison rather than Ethel or Linda, wear their baseball caps backwards, raise their pitch at the end of a sentence? The process is still poorly understood by social science, with its search for external causes of behavior, but is essential to bridging the largest chasm in intellectual life: that between individual psychology and collective culture."[13]

As above, so below. The effort to make things on different levels conform to the same rules had become explicit. Economics writer and *The Wisdom of Crowds* author James Surowiecki explained to libertarian *Reason* magazine that Hayek's notion of catallaxy—amplified by Santa Fe's computers—could well become a universal approach to understanding humans: "In the 20th Century, this insight helped change the way people thought about markets. In the next century, it should change the way people think about organizations, networks, and the social order more generally."[14]

And so scientists, economists, cultural theorists, and even military strategists[15] end up adopting fractalnoia as the new approach to describing and predicting the behavior of both individual actors and the greater systems in which they live. Weather, plankton, anthills, cities, love, sex, profit, society, and culture are all subject to the same laws. Everything is everything, as Bateson's theory of Mind finds itself realized in the computer-generated fractal.

Where all these scientists and social programmers must tread carefully, however, is in their readiness to draw congruencies and equivalencies between things that may resemble one another in some ways but not others. Remember, the fractal is self-similar on all its levels, but not necessarily identical. The interactions between plankton in the coral reef may be very similar to those between members of Facebook—but they are not the same. Drawing too many connections between what's happening on the molecular level and what's happening on a social or intellectual level can be hazardous to everyone involved.

Not even the economists who came up with these models are

particularly immune from their often-imprecise predictions and recommendations. Many of the "quant" teams at hedge funds and the risk-management groups within brokerage houses use fractals to find technical patterns in stock market movements. They believe that, unlike traditional measurement and prediction, these nonlinear, systems approaches transcend the human inability to imagine the unthinkable. Even *Black Swan* author Nassim Taleb, who made a career of warning economists and investors against trying to see the future, believes in the power of fractals to predict the sudden shifts and wild outcomes of real markets. He dedicated the book to Benoit Mandelbrot.

While fractal geometry can certainly help us find strong, repeating patterns within the market activity of the 1930s Depression, it did not predict the crash of 2007. Nor did the economists using fractals manage to protect their banks and brokerages from the systemic effects of bad mortgage packages, overleveraged European banks, or the impact of algorithmic trading on moment-to-moment volatility.

More recently, in early 2010, the world's leading forecaster applying fractals to markets, Robert Prechter, called for the market to enter a decline of such staggering proportions that it would dwarf anything that has happened in the past three hundred years.[16] Prechter bases his methodology on the insights of a 1930s economist, Ralph Nelson Elliott, who isolated a number of the patterns that seem to recur in market price data. They didn't always occur over the same timescale or amplitude, but they did have the same shape. And they combined to form larger and larger versions of themselves at higher levels, in a highly structured progression.

Prechter calls this progression the Elliott Wave. We may as well call it fractalnoia. For not only is the pattern supposed to repeat on different scales and in progressively larger time frames; it's also supposed to be repeating horizontally across different industries and aspects of human activity. Prechter titled his report on fractals and the stock market "The Human Social Experience Forms a Fractal." And, at least as of this writing, the biggest market crash since the South Sea Bubble of 1720 has yet to occur.

Fractalnoia of this sort is less dangerous for the individually incorrect predictions it suggests to its practitioners and their customers than for the reductionist outlook on humanity that it requires. Machines, math equations, molecules, bacteria, and humans are often similar but hardly equivalent. Yet the overriding urge to connect everything to everything pushes those who should know better to make such leaps of logic. To ignore the special peculiarities, idiosyncrasies, and paradoxes of activity occurring on the human and cultural level is to ignore one's own experience of the moment in order to connect with a computer simulation.

TO BE OR TO BE

Ironically perhaps, the way past the problem of human unpredictability may be to work with it and through it rather than to ignore it. Fractals may be generated by computers, but the patterns within them are best recognized and applied by people. In other words, pattern recognition may be less a science or mathematic than it is a liberal art. The arts may be touchy-feely and intuitive, but in eras of rapid change such as our own, they often bring more discipline to the table than do the sciences.

When I was exposed to computers for the very first time, back in the mid-1970s, I remember feeling like quite the artist among geeks. Everyone else at my high school's computer lab was already a devoted mathematician. I was a theater enthusiast. The programs they wrote on our IBM terminals had to do with proving the Pythagorean Theorem or utilizing Planck's Constant. I was more interested in making pictures out of letters, or getting the whole machine to seize up with what I considered to be "performance art" programs such as:

```
10: Print "Eat Me"
20: Escape = "off"
30: Go to 10
```

Understandably, I was not appreciated, and so I abandoned technology for the arts.

After I graduated from college with a theater degree in the mid-1980s, I learned that the most artsy and psychedelic folks with whom I went to school had gone off to Silicon Valley to become programmers. It just did not compute. So I flew out to California to find out what had compelled these Grateful Dead–listening hashish eaters to get so straitlaced and serious. Of course, it turns out they weren't straight at all. They were building cyberspace—a task that required people who were not afraid to see their wildest dreams manifest before their own eyes. They programmed all day (and scraped the buds off peyote cactuses at night).

I interviewed dozens of executives at companies such as Northrup, Intel, and Sun and learned that they were depending upon these legions of psychedelics-using "long hairs" to develop the interfaces through which humans would one day be interacting with machines and one another. "They already know what it's like to hallucinate," one executive explained. "We just have to warn them before the urine tests, is all." Silicon Valley needed artistic visionaries to bring the code to life, and so they tapped the fringes of arts culture for them.

Now that more than twenty years have passed, it is the artists who are trying to keep up with the technologists. At digital arts organizations such as Rhizome and Eyebeam, as well as digital arts schools from NYU's Interactive Telecommunications Program and Parsons Design and Technology program, it is the technologists who are leading the way. Self-taught, long-haired programmers create new routines and interfaces, while well-dressed and well-read MFAs struggle to put this work into historical context. The artists are the geeks, and the programmers are the performers. The programmers conceive and create with abandon. The artists try to imagine what humans may want to do with these creations, as well as what these creations may want to do with *us*.

Indeed, the more technologized and interconnected we become, the more dependent we are on the artist for orientation and pattern recognition. While I strongly advocate the teaching of computer programming to kids in grade school, I am just as much a believer in teaching kids how to think critically about the programmed environments in which they will be spending so much of their time. The former is engineering; the latter is liberal arts. The engineers write and launch the equations; the liberal artists must judge their usefulness, recognize the patterns they create, and—oh so very carefully—generalize from there. For the artist—the human, if you will—this care comes less from the accumulation of more and more specific data than the fine-tuning of the perceptual apparatus. In a fractal, it's not how much you see, but how well you see it.

In evaluating the intelligence failures surrounding the US overthrow of Iraq, University of Pennsylvania psychologist and decision theorist Philip Tetlock concluded that people with more wide-ranging and less specific interests are better at predicting the future than experts in the very fields being studied.[17] Using Isaiah Berlin's prototypes of the fox and the hedgehog, Tetlock studied hundreds of different decisions and determined that foxes—people with wider and less specific interests—saw patterns more accurately than hedgehogs, who used their knowledge of history and their prior experiences to predict future outcomes.

Hedgehogs, Tetlock learned, were too eager for "closure" to be able to sit with the uncertainty of not knowing. They attempted to use their fixed ideas to determine where things were going. Foxes had less at stake in conventional wisdom, so they were much more likely to use probability in making their assessments. While hedgehogs were locked into their own disciplines, foxes could think laterally, applying the insights from one field to another. Knowledge had handicapped the hedgehogs, while the wide-ranging curiosity of the foxes gave them the edge.

Now, on the surface this sounds like the failing of the fractalnoids—those economists who want to equate the properties of plankton with the personalities of Parisians. But it's fundamentally different in that it's human beings applying patterns intuitively to different systems, not the frantic confusion of apples and oranges or, more likely, apples with planets. Yes, it is still fraught with peril, but it's also a rich competency to develop in an era of present shock.

For instance, I still don't know whether to be delighted or horrified by the student who told me he "got the gist" of *Hamlet* by skimming it in a couple of minutes and reading a paragraph of critique on Wikipedia. The student already saw the world in fractal terms and assumed that being able to fully grasp one moment of *Hamlet* would mean he had successfully "grokked" the whole. For him, the obvious moment upon which to seize was that "superfamous one, 'to be or not to be.'" The student went on to explain how everything about the play extended from that one point: Hamlet can't decide whether to take action or whether to kill himself. But he's not even really contemplating suicide at all—he's only pretending to do this because he's being watched by Polonius and the man who murdered his father. "It's all action and inaction, plays within plays, and the way performance and identity or thought and deed are confused," the student explained to me. "The readiness is all. The play's the thing. I get it. Trust me."

Trust him? It's hard not to—particularly when I take into account that he's living in a world where he's expected to accomplish more tasks each minute than I did in an hour or a whole day at his age. Functioning in such a world does require getting the "gist" of things and moving on, recognizing patterns, and then inferring the rest. It's the sort of intellectual fakery my college friend Walter Kirn describes in the novelized account of his education, *Lost in the Meritocracy*, in which the protagonist succeeds brilliantly at Princeton by faking it. He explains with a mix of pride and self-contempt:

> I came to suspect that certain professors were on to us, and I wondered if they, too, were actors. In classroom discussions, and even when grading essays, they seemed to favor us over the hard workers, whose patient, sedimentary study habits were ill adapted, I concluded, to the new world of antic postmodernism that I had mastered almost without effort.[18]

What may have begun as a fakery, though, became a skill itself— a way of recognizing patterns and then surfing through a conversation in an increasingly convoluted, overlapping, and interdisciplinary academic world. Kirn graduated summa cum laude with a scholarship to Oxford and became a respected author and critic. This wide-angle approach may be not the only skill one needs to meet intellectual challenges, but it's as crucial to understanding and performance as is focused study. The truly accomplished musician can do more than play his repertoire; he can also pick up a "fake book" and instantly play one of thousands of tunes by scanning a melody and chord chart. He knows enough of the underlying patterns to fill in the rest.

This more generalist and intuitive perspective—the big view, if you will—has been studied extensively by University of Michigan psychologist Richard Nisbett. Like Tetlock, Nisbett found that inductive logic was being undervalued in decision making and that people's reasoning would be vastly improved by weighing the odds and broadening focus, rather than just relying on highly defined expertise and prior experiences.

After noticing how differently his Chinese graduate students approached certain kinds of problems, Nisbett decided to compare the ways East Asians and Westerners perceive and think. He showed students pictures of animals in their natural environments and tracked their eye movements as they scanned the images. While his American students invariably looked at the animal first, and only then took time to look at the background, his Asian students looked at the forest or field first. When they did finally look at the tiger or elephant, they spent much less time on it than their American counterparts. When asked about the pictures later, the American students had much better recall of the specific objects they had seen, but the Chinese could recount the background in great detail. Nisbett could even fool his Asian students into believing they hadn't seen an image before, simply by changing the background. He did the same thing to the Americans: when he completely changed the environment but left the subject animal alone, they had no idea the picture had changed. They were fixated on the focal point and blind to the greater environment.

Westerners tended to focus on objects and put them into categories, while Easterners looked at backgrounds and considered bigger environmental forces. Westerners used formal logic to figure things out, while Easterners used a variety of strategies. As Nisbett explains, "East Asians reason holistically—that is, they focus on the object in its surrounding field, there is little concern with categories or universal rules, and behavior is explained on the basis of the forces presumed to be operative for the individual case at a particular time. Formal logic is not much used and instead a variety of dialectic reasoning types are common, including synthesis, transcendence and convergence."[19]

These pairs—the American and the Asian, the hedgehog and the fox, the expert and the generalist—suggest two main ways of managing and creating change: influence the players or manipulate the greater environment. When we focus on the pieces, we are working in what could be considered the timescape of *chronos*: we break objects down into their parts, study them carefully, focus in, and engage with the pieces scientifically. Like dissection, it works best when the subjects are dead. When we focus on the space around the pieces, we have shifted to the time sensibility of *kairos*. The space between things matters more than the things themselves. We are thinking less about tinkering with particular objects than about recognizing or influencing the patterns they create and the connections they make. We stop getting dizzy following the path of every feedback loop and pull back to see the patterns those loops create.

In a more practical sense, it's the difference between trying to change your customer's behavior by advertising to him, or changing the landscape of products from which he has to choose; trying to convince people in a foreign nation to like you by crafting new messaging, or simply building a hospital for them; planning by committee where to pave the paths on a new college campus, or watching where the grass has been worn down by footsteps and putting the paths there.

As economist and policy consultant Joshua Ramo puts it, focusing on the environment gives one access to the "slow variables" that matter so much more in the long run than surface noise. His book *The Age of the Unthinkable* is directed primarily at policy makers and military strategists, but his insights apply widely to the culture of the fractal. In language likely surprising to his own community of readers, Ramo suggests leaders develop "empathy." He doesn't mean just crying at the misfortune of others but rather learning how to experience the world through the sensibilities of as many other people as possible.

Ramo is particularly intrigued by a Silicon Valley venture capitalist, Sequoia Capital partner Michael Moritz, a very early and successful investor in Google and YouTube. Unlike most of his peers, Moritz did not break down companies or technologies in order to understand them. "Moritz's genius was that he wasn't cutting companies up as he looked at them but placing them in context. The forces shaping a technology market, consumer demands, changes in software design, shifts in microchip pricing, the up-and-down emotions of a founder—he was watching all of these for signs of change."[20]

Moritz understood Google's single-pointed vision to index the Web but insisted that the only path to that goal was to constantly improvise. In order to lead his companies through the kinds of changes they needed to make, he believed he needed to understand their goals as well as the founders did. He required the ability to lose himself in their dreams. "The thing I am terrified of is losing that empathy," Moritz explains. "The best investments we have missed recently came because the founders came in here and we blew them off because we didn't understand them. We couldn't empathize. That is a fatal mistake. If I were running American foreign policy, I would want to focus on empathizing."[21]

Empathy isn't just studying and understanding. It's not something one learns but a way of feeling and experiencing others. It's like the difference between learning how to play a song and learning how to resonate with one that's already being played. It is less about the melody than the overtones. Or in networking terms, less about the nodes than the connections between them.

My own Zen master of such connections is Jerry Michalski, a technology analyst who used to edit Esther Dyson's *Release 1.0* newsletter and then left to go into what seemed to me to be a sort of public reclusiveness. I'd see him at almost every technology or digital culture conference I could make time to attend, usually because I was doing a talk or panel myself. Jerry seemed content to attend for attendance' sake. He'd just sit there with his laptop, tapping away during everyone's speeches, actually reading the PowerPoint slides people put up behind them, and, later, Tweeting quotes and re-Tweeting those quoted by others.

I first met Jerry when I sat next to him at a conference in Maine in 1998, shortly after he had resigned from his editorial post. We were listening to someone I don't remember speak about something I can't recall. What I do remember is that I peeked down to see what he was writing on his computer. It wasn't text at all, but a visualized web of connections between words. "It's called TheBrain,"[22] Jerry told me. He would enter a name, a fact, a company—whatever. And then he'd connect it to the other things, people, books, ideas, in his system already. He put me in, added my speaking and literary agents, connected me to my books, my influences, my college, as well as some of the ideas and terms I had coined. "I think it might be useful for understanding what's going on."

As of today, Jerry has over 173,000 thoughts in his brain file, and 315,000 links between them, all created manually.[23] And he is finally emerging from his public reclusiveness to share what he learned with the rest of us. His chief insight, as he puts it, is that "everything is deeply intertwingled." He no longer thinks of any thought or idea in isolation, but in context. Jerry describes himself as a pattern finder and lateral thinker. Everything matters only insofar as its relationship to everything else.

Surprisingly, offloading his memory to the computer in this fashion has not damaged his recall, but improved it. The act of storing things in relationship to other things has reinforced their place in his real brain's memory. And going back into TheBrain "refreshes my neural pathways." They are not random facts but parts of the greater fabric—components in the greater pattern.

Most important, however, Michalski has become a great advocate for what he calls "The Relationship Economy"—the economy of how people are connected to one another. As he and microfinance innovator April Rinne explained in a recent *Washington Post* op-ed, "Say, for example, you are trying to solve a complex problem such as the global financial crisis. Do you ask an economist, a sociologist or a political scientist? Each of them individually is too constrained. The more multi-faceted the problem, the more forces intersect and the more challenges one must face within a siloed system."[24]

On the one hand, this means democratizing innovation and change—not crowdsourcing, per say, where you just get customers to do your advertising work for you—but creating an environment where everyone connected with a culture or industry feels welcome to participate in its development. It's the way amateur gearheads develop the best new technologies for cycling, which are then incorporated into the designs and products of major manufacturers. It's the way Adobe encourages the users of its programs to create their own plug-ins, which are then shared online or incorporated into the next release. It's the way Google provides free online tutorials on how to make Android apps, in the hope that an open development culture can eventually beat Apple's "walled garden" of more professionally devised products.

On the other hand, it means coming to understand that—at least in the fractal—one's relationships matter more than one's accumulated personal knowledge; the shared overtakes the owned; connections supersede the ego. Contrast, for example, Michalski's selfless Brain to Microsoft's celebration of individual immortality online, the MyLifeBits project. In this public display of solipsism, we are encouraged to marvel at and model the behavior of researcher Dr. Gordon Bell, who has scanned, recorded, and uploaded everything he can about himself—lectures, memos, letters, home movies, phone calls, voicemail—since 1998.[25] The reward for turning his entire life into bits is that Microsoft's archival software will then allow him to access anything he has done or seen, at any time in his life.

The promise is that, through computers, we gain a perfect memory. Bell and Microsoft claim they are realizing Vannevar Bush's original 1950s dream for personal computing, which was to externalize one's memory into a perfect record: total recall. In reality, of course, the project models and publicizes a mode of behavior that would make a market researcher drool. This is the Facebook reality, in which we operate under the false assumption that we are the users of the platform, when we are actually the product being sold. Moreover, on a subtler level, it uses computers to heighten the sense that we are what we perceive and experience as individuals, and that the recordable bits of information about ourselves reflects who we are. The marketer loves data on individuals, since it's individuals who can be influenced to make purchases. The market loves individuals, too, since the less networked people are, the less they share and the more they have to buy for themselves.

More important, however, as an approach to a postnarrative world in which living networks replace linear histories, MyLifeBits is entirely more vulnerable to fractalnoia than TheBrain. For Michalski, the "self" is defined by connections. TheBrain isn't about him, but about everyone else he has met. It is not limited to a single path, but instead suggests an infinity of possible pathways through data. The patterns are out there, constantly evolving and becoming increasingly "intertwingled." It's all potential energy. Maintaining this state—this readiness-is-all openness—isn't easy, but it's the only way to create context without time.

MyLifeBits, on the other hand, approaches memory as the storage of a personal narrative over time. It is not really stored potential energy as much as a record of spent kinetic energy. It is a diary, both egocentric and self-consumed. Moreover, once stored, it is locked down. History no longer changes with one's evolving sensibilities; it describes limits and resists reinterpretation. One's path narrows. As far as doing pattern recognition in a landscape of present shock, the user must identify entirely with sequences of self. Everything relates, as long as it relates back to himself. Where was I when I saw that? How did I think of that the first time I did it? How does that reflect on me?

This is the difference between the networked sensibility and paranoia—between pattern recognition and full-fledged fractalnoia. The fractalnoid is developing the ability to see the connections between things but can only understand them as having something to do with himself. This is the very definition of paranoia. As the title of Gordon Bell's best archived speech on the project explains it, "MyLifeBits: A Transaction Processing Database for Everything Personal."

But that's just it: in a fractal landscape, *nothing* is personal. This may be the hardest lesson for victims of present shock to accept: it's not about you.

CHAPTER 5

APOCALYPTO

As I saw it, his big mistake was showing me the compound at all. But I suppose the numerous construction crews and contractors responsible for converting this former missile silo into an apocalypse bunker know its location as well. So does the documentary crew from History Channel who filmed him there a few weeks before my own visit. Still, a promise is a promise, so I won't tell you in which Midwestern state (starts with *K*) "Dan" (not his real name) has built the bunker he believes will be capable of sustaining him and his family through the apocalypse.

"I don't mean apocalypse in the religious way," Dan explains as he escorts me down the single spiral metal staircase leading to the living quarters. (Hard to get out in a fire, I suppose, but easier to defend if attacked.) "I'm thinking *Contagion*, *Asteroid*, even *China Syndrome*," he explains, using movie-title shorthand for global pandemic, collision with an asteroid, or nuclear meltdown. He is used to being interviewed.

"What about *The Day After Tomorrow?*" I suggest, bringing climate change into the mix.

"Not likely," Dan says. "That's been debunked."

Dan is a former real estate assessor who now makes his living selling information online to "preppers" like himself, who have assumed that catastrophe is imminent and that the best way through is to prepare for the inevitable collapse of civilization as we know it. Even without an apocalypse, the bunker is about as nice as one could expect a windowless underground apartment to get. It has been built into the control rooms of a decommissioned nuclear missile silo that is surrounded by a dozen feet of concrete in all directions.

Still, the air is significantly cooler and more pollen-free than the summer-scorched fields aboveground, making the place feel more than livable. There's a little kitchen done up in '70s colors: lime green laminate countertops and orange vinyl chairs. It's a bit like what you'd see in a nice trailer home or on a houseboat. There's a door leading to a series of pantries I'm told contain a ten-year supply of food for six people. On the next level, three bedrooms and a media room with built-in monitors and a couch that looks like it time traveled from an *Avengers* episode. The whole place has been meticulously designed—not just its interior finishes but the solar-powered generators, the air-purification system, the radiation shielding, and the intrusion-deterrence system. It is as well thought-out as ten screenplays, with each doomsday scenario integrated into the layout or hardware.

I can't help but imagine myself in this setting, having time to watch the entire Criterion Collection of DVDs, read the philosophy I've never had time for (most of whatever happened between Saint Thomas Aquinas and, say, Francis Bacon), or get to know my family without the pressures of homework, the Internet, or neighbors. Time for . . .

And then I realize I have been sucked in by the allure. This missile silo repurposed as a bomb shelter isn't a Plan B at all, but a fantasy. Whether Dan ever has to—or *gets* to—live in this place, its mere creation may be its truest purpose. Where the basement model railroad once gave the underachiever a chance to build and run a world, the doomsday apartment gives the overwhelmed present-shock victim the chance to experience the relief of finality and a return to old-fashioned time.

Dan and thousands of other preppers and doomsdayers around the world (but particularly in the United States) expect a complete societal breakdown in their own lifetimes. Their visions are confirmed by media such as History Channel's *Armageddon* week or ads on Christian and right-wing radio for MREs (meals, ready to eat, used by soldiers) and silver coins (to use after the collapse of the banking system). A company named Vivos is selling reservations for apartments in a Walmart-sized bunker in Nebraska. An initial down payment of $25,000 earns you a place in an underground community where they have thought of everything, from a beauty salon to a small prison for those who might get out of control after the world is gone.

Sales of Vivos and similar bunkers increase tenfold during well-covered disasters such as nuclear accidents, pandemic scares, and terrorist events. But there's more than rational self-preservation at work here. Apocalyptic headlines give justification to a deeper urge. Any natural or man-made disaster simply provides the pretense to succumb to what we will call *apocalypto*—a belief in the imminent shift of humanity into an unrecognizably different form.

At least the annihilation of the human race—or its transmogrification into silicon—resolves the precarious uncertainty of present shock. So far in our journey, we have seen the human story collapse from a narrative into an endless occupation or infinite game. We have seen how digital technology continually challenges our coherence and connection to the natural rhythms that used to define our biology and psychology alike. We have watched banks and businesses compress time into time, leveraging the moment like an overwound spring. And we have seen identity itself devolve into a nonlocal pattern in a depersonalized fractal. Apocalypto gives us a way out. A line in the sand. An us and a them. And, more important, a before and an after.

That's why it's important that we distinguish between valid concerns about the survival of our species and these more fantastic wishes for reversal and recognition—the story elements at the end of all heroic journeys. If anything, the common conflation of so many apocalypse scenarios—bird flu, asteroid, terrorist attack—camouflages ones that may actually be in progress, such as climate change or the slow poisoning of the oceans. In their book *The Last Myth*, Mathew Barrett and Mel Gilles put it this way:

> By allowing the challenges of the 21st century to be hijacked by the apocalyptic storyline, we find ourselves awaiting a moment of clarity when the problems we must confront will become apparent to all—or when those challenges will magically disappear, like other failed prophecies about the end of the world. Yet the real challenges we must face are not future events that we imagine or dismiss through apocalyptic scenarios of collapse—they are existing trends. The evidence suggests that much of what we fear in the future—the collapse of the economy, the arrival of peak oil and global warming and resource wars—has already begun. We can wait forever, while the world unravels before our very eyes, for an apocalypse that won't come.[1]

For many, it's easier, or at least more comforting, to approach these problems as intractable. They're just too complex and would involve levels of agreement, cooperation, and coordination that seem beyond the capacity of humans at this stage in our cultural evolution, anyway. So in lieu of doing the actual hard work of fixing these problems in the present, we fantasize instead about life afterward. The crisis of global warming morphs into the fantasy of living off the grid. The threat of a terrorist attack on our office tower leads us to purchase an emergency personal parachute for easy egress, and to wonder how far up the org chart we might be promoted once everyone else is gone. The collapse of civilization due to nuclear accident, peak oil, or SARS epidemic finally ends the ever-present barrage of media, tax forms, toxic spills, and mortgage payments, opening the way to a simpler life of farming, maintaining shelter, and maybe defending one's family.

The hardest part of living in present shock is that there's no end and, for that matter, no beginning. It's a chronic plateau of interminable stresses that seem to have always been there. There's no original source to blame and no end in sight. This is why the return to simplicity offered by the most extreme scenarios is proving so alluring to so many of us.

I, ZOMBIE

Even those of us who aren't storing up on survival shelter supplies at Costco (yes, they now sell MREs and other apocalypse goods) are nonetheless anxious to fantasize about the coming Armageddon. In popular culture, this wish fulfillment takes the form of zombie movies and television shows, suddenly resurrected in the twenty-first century after decades of entombment. AMC's *The Walking Dead*, in which a ragtag group of fairly regular folks attempts to survive a total zombie apocalypse, is the highest-rated basic-cable drama of all time.

Like any great action drama, a zombie show gives its viewers the opportunity to strategize, vicariously, on a very simple playing field. Scenario: two people are running from a horde of flesh-eating zombies and losing ground. If one person shoots the other, will this attract enough zombies to the victim to allow the shooter to get away? Is such a sacrifice ethical? What if it allows the shooter to return to the camp with medical supplies that save a dying child?

This was the climax of just one episode of *The Walking Dead*, which—like many others—spawned countless pages of online discussion.

The Prisoner's Dilemma–like clarity of the scenario reassures modern audiences the way Cain and Abel simplified reality for our ancestors. But in our case, there's no God in judgment; rather, it's the zero-sum game of people with none of civilization's trappings to mask the stark selfishness of every choice, and no holy narrative to justify those choices.

The zombie legend originated in the spiritual practices of Afro-Caribbean sects that believed a person could be robbed of his soul by supernatural or shamanic means and forced to work as an uncomplaining slave. Canadian ethnobotanist Wade Davis studied Haitian voodoo rituals in the 1980s and determined that a kind of "zombie" state can be induced with powerful naturally derived drugs. In horror films, people become zombies by whatever process is deemed scariest by the filmmaker of the era—magic, possession, viral infection—but the result is the same. The victim becomes a walking corpse, a human without a soul.

Indeed, zombies are the perfect horror creations for a media-saturated age in which we are overloaded with reports of terrorism, famine, disease, and warfare. Zombies tap into our primal fear of being consumed and force us to come up with something—anything—to distinguish ourselves from the ever-hungry, animated corpses traipsing about the countryside and eating flesh. Deep down, these schlocky horror flicks are asking profound questions: What is life? Why does it depend on killing and consuming other life? Does this cruel reality of survival have any intrinsic meaning? How will it all end?

The way in which zombie movies pose these questions has changed significantly over time, telling us more about ourselves, and about what we most fear, in the process. Zombies have been a staple of American filmmaking since the indie flick *White Zombie* (1932), best remembered for its eerie shots of undead slaves staring into the night. In that movie, Bela Lugosi plays an evil sorcerer who promises to turn a woman into a zombie so that her spurned lover can control her forever, presumably as a mindless sex servant. Perfect fare for a nation finally reckoning with its own population of former slaves, as well as one of preliberated females just beginning to find their own voices. Back then, though, the big questions seemed to have more to do with whether a walking dead servant or wife could fully satisfy a man's needs. (Given the outcome, apparently not.) By 1968 George Romero's low-budget classic *Night of the Living Dead* had reversed this dynamic. Now it was up to the film's human protagonists to distinguish themselves from the marauding bands of flesh eaters—and to keep from being eaten. Racial conflicts among the film's living characters end up costing them valuable time and resources; against the backdrop of attacking zombies, the racial tension of the late 1960s seems positively ludicrous. The film's African American hero survives the night but is mistaken for a zombie and shot dead the next morning.

The film's sequels had survivors holing up in places like shopping malls, through which zombies would wander aimlessly all day as if retracing the steps of their former lives as consumers. Of course, the real consumption begins when the zombies find humans on whom to feast—an irony not lost on one tough guy who, as his intestines are being eaten, has enough wit to shout, "Choke on 'em!" What makes the humans for whom we're rooting any different from the zombies by whom we're repulsed? Not much, except maybe cannibalism, and the technical distinction that our humans are living while the zombies are "living dead."

State-of-the-art zombie films—most notably *28 Days Later* (2002) and its sequel, *28 Weeks Later* (2007)—use the undead to explore today's hazier ethical climate. Instead of fearing magic or consumerism, we are scared of the unintended consequences of science and technology. Perhaps that's why rather than reaching zombification through magic or rampant consumerism, the undead in this film series have been infected by a man-made virus called "rage."

Playing to current apocalyptic fears, the zombies in *28 Days Later* wipe out the entirety of England, which has been quarantined by the rest of the world in a rather heartless but necessary act of self-preservation. Like the hilarious but unironically fashioned book *The Zombie Survival Guide* (2003), here's a zombie tale for the 9/11 era, when fantasies of urban chaos and duct tape–sealed apartment windows are no longer relegated to horror films; these paranoid scenarios became regular fare on CNN.

In *28 Weeks Later*, well-meaning American troops work to rebuild England by putting survivors in a protected green zone and even firebombing the innocent in a desperate attempt to quash a zombie insurgency. The movie's undead ruthlessly attack anyone for flesh, and its weaker characters choose to save their own skins instead of protecting their wives and children. The film's heroes distinguish themselves and redeem our view of humanity through acts of self-sacrifice. It turns out, however, that they've sacrificed themselves on behalf of a child who carries the virus and goes on to infect the rest of the world. Humanity, like civil liberty, is no longer a strength but a liability.

In the TV series *The Walking Dead*, as well, we are to question who truly are the ones who have lost their humanity—whatever that may have been. In the season three finale (the most-watched basic-cable hour ever) the protagonist murders his best friend and squad car partner—who just so happens to be in love with his wife. The writers are at pains to cast the humans as not simply responding as necessary to apocalyptic circumstances, but using these circumstances as an excuse to act upon long-repressed impulses. The zombie apocalypse not only relieves us of our highly stressed, over-civilized, and technologically determined lives, it reveals the savagery and selfishness innate to our species. We have no morality separating us from brute nature or even lifeless matter, so we humans may as well be walking dead.

TRANSCENDING HUMANITY

For all the flesh eating going on in the zombie genre, there's something positively flesh loathing about the psychology underlying it. People are the bad guys. Apocalypto seems less about transforming the human species than transcending it altogether. In neither the hallucinations of psychedelic 2012 end-of-worlders nor the scenarios forecast by techno-enthusiast extropians do we humans make it through the chaos attractor at the end of time—at least not in our current form. And why should we want to, when human beings are so loathsome, smelly, and inefficient to begin with? The postnarrative future belongs to the godhead, machines, cockroaches, planetary intelligence, complexity, or information itself.

Of course, it's not really the future, since—according to a good many of these apocalyptans—time will have stopped entirely. To them, present shock is not a metaphor at all—not a state of confusion or a dynamic between people and their increasingly presentist society—but rather an existence outside time. That's part of what makes it so fantastic to think about, but also so inhospitable to cellular organisms such as you and me.

To understand this strain of present shock, we have to go back to the fractal and the way it inspires people to look for patterns. My first exposure to the logic of apocalypto came through an old friend of mine, a shamanic explorer with a penchant for Irish folklore named Terence McKenna. Terence, one of the most articulate stoned heroes of the psychedelics underground, saw in fractals a way to pattern time itself. Back in the 1970s, he and his brother Dennis spent several months in the Amazonian rainforest, ingesting native mushrooms and other potent psychedelics. During these vision quests, the brothers experienced themselves traveling out of body, inside the body, and through the core of human DNA. After a particularly harrowing excursion in which one of the brothers got "lost" between dimensions, Terence became obsessed with navigating this timeless terrain. He wanted to make sense of the infinity of the fractal.

He emerged with a new understanding of time as having an endpoint—a "teleological attractor," as Terence put it—drawing us toward greater interconnectedness and complexity. The increasing intensity of our era can be attributed to our nearing the event horizon of this attractor. It's basically like a waterfall or black hole in the time-space continuum that we are drawn toward, fall through, and are then utterly changed—if we make it out the other end at all. According to McKenna's schema, things keep getting more and more complex, interconnected, and unbearably strange, like a really weird and scary acid trip, where everything becomes part of the same pattern. Once everything ends up connected to everything else, reality itself reaches a singularity—a moment of infinite complexity in which everything occurs simultaneously. It's a moment of absolute present shock, in which history and the future and present fold into one another, ending time altogether.

Terence immediately set upon figuring out just when this might happen and ended up using the sequence of the *I Ching*—the Taoist Book of Changes and divination system—as the basis for a numerical formula that maps the rise and fall of novelty over any period of time. McKenna's Timewave Zero, as he calls it, is a shape—a linear graph— that is to be overlaid on the time line of history in order to figure out when things get weirder or less so. With a little bit of rejiggering, McKenna was able to lay out his zigzaggy repeating fractal pattern in such a way that the biggest period of the graph ended right on December 21, 2012—the same day purported to be indicated by the Mayan calendar as the end of time.

For McKenna, who ended up dying before the prophesied end date, the increasing "novelty" in the world—from war and market crashes to disease and environmental disaster—are not signs of death but of birth. He often remarked that were a person who knew nothing of human biology to come upon a woman giving birth, he would think something terribly wrong was going on. She would appear to be dying, when she was actually giving life to a new human being. Such is the state of a civilization on the precipice of the singularity.

The problem is, not everyone makes it through the attractor at the end of time. According to McKenna, only those who can successfully navigate the all-at-onceness of a posthistorical reality will be able to make sense of existence there. By inference, the object of the game is to do enough strong psychedelics now so that you'll know how to navigate a landscape as precarious as the one a person visits while hallucinating on the rainforest psychedelic DMT (dimethyltryptamine).

One person who surely qualifies is author and spiritual teacher Daniel Pinchbeck, whose own interactions with the Amazonian vision plants convinced him that he could hear the voice of the Mayan god Quetzalcoatl. The plumed serpent warned Pinchbeck of humanity's abuse of the planet, while also confirming the imminent shift beyond time, scheduled for 2012. The signs of the next age are everywhere: crop circles, ESP experiments, particle physics, UFOs, time travelers, and so on. While Pinchbeck's global transformation may also be an exclusive event, limited to those who "get it," at least he has dedicated himself and his network of followers to mitigating the impact of this shift on the rest of us. Pinchbeck is an advocate of permaculture farming, local currencies, and other techniques through which we may combat the materialism currently guiding human activity.

Again, though, making it through the attractor at the end of time requires more than mere compassion or a willingness to work together. We must abandon individuality altogether and accept our place in the new cosmic order. We surrender our illusion of distinctness and admit that we are part of nature. We sacrifice Father Time to return to Mother Earth. Human progress has been a sham—a painful, costly, and destructive detour, or, at best, a necessary stage in our release from the shackles of matter altogether.

Such narratives find their origins in the writings of theologians such as Pierre Teilhard de Chardin, the early-twentieth-century French Jesuit priest and paleontologist who saw human beings evolving toward an Omega Point of supreme consciousness. Just as cells joined up and evolved into organisms, we humans will evolve together into a greater single being. It's a nice image, and one I've contemplated on numerous occasions, but not a stage of evolution that feels particularly imminent—no matter how many Facebook friends I happen to accumulate or how overwhelmed I become by the virtual connections.

IT'S THE INFORMATION, STUPID

None of this increasing complexity would be a problem if it weren't for our darned human limits. That's why the latest breed of apocalyptans— an increasingly influential branch of the digerati who see technology as the true harbinger of the singularity—mean to help us accept our imminent obsolescence. Echoing the sentiments of the ancient ascetic, they tend to regard the human physical form with disregard or even disdain. At best, the human body is a space suit for something that could be stored quite differently.

The notion of a technologically precipitated singularity was popularized by futurist and electronic music engineer Ray Kurzweil. In his book *The Age of Spiritual Machines*, Kurzweil argues that human beings are just one stage in the evolution of matter toward higher levels of complexity. Yes, cells and organisms are more complex than mere atoms and molecules, but the human capacity for continuing development pales in the face of that of our machines. The very best thing we have to offer, in fact, is to continue to service and develop computers until the point—very soon—when they are better at improving themselves than we are. From then on, technological evolution will outpace biological evolution. Whatever it is that makes us uniquely human, such as our genome or cognitive functioning, will have been mapped and virtualized by computers in around 2050, anyway. We may as well stand aside and let it rip.

Thanks to Kurzweil's Law of Accelerating Returns, technology develops exponentially and has been doing so since time began. But it is only getting really interesting now that we have rounded the bend of the exponential curve to a nearly vertical and infinite shot upward. The antithesis of the Law of Diminishing Returns, the Law of Accelerating Returns holds that technology will overtake humanity and nature, no matter what. In his numerous books, talks, and television appearances, Kurzweil remains unswerving in his conviction that humanity was just a temporary step in technology's inevitable development.

It's not all bad. According to Kurzweil, by 2029 artificial intelligences will pass the Turing test and be able to fool us into thinking they are real people. By the 2030s, virtual-reality simulations will be "as real and compelling as 'real' reality, and we'll be doing it from within the nervous system. So the nanobots in your brain—which will get to your brain through the bloodstream, noninvasively and without surgery—will shut down the signals coming from your real senses and replace them with senses that your brain will be receiving from the virtual environment."[2] Just be sure to read the fine print in the iTunes agreement before clicking "I agree" and hope that the terms don't change while you're in there.

Slowly but surely, the distinction between our real memory or experiences and our virtual ones ceases to have any meaning. Eventually, our nonbiological mechanisms take over where our biological ones leave off. Consciousness, such as it is, is better performed by some combination of microchips and nanobots than our old carbon sacks, and what we think of as people are discontinued.

Kurzweil may push the envelope on this line of thought, but a growing cadre of scientists and commentators have both wittingly and unwittingly gotten on his singularity bandwagon. Their credentials, intelligence, and persuasiveness make their arguments difficult to refute.

Kevin Kelly, for instance, convincingly portrays technology as a partner in human evolution. In his book *What Technology Wants*, he makes the case that technology is emerging as the "seventh kingdom of life on Earth"—along with plants, insects, fungi, and so on. Although he expresses himself with greater humility and admirable self-doubt than Kurzweil, Kelly also holds that technology's growth and development is inevitable, even desirable. Yes, certain technologies create problems, but that just opens the opportunity for yet another technology to mitigate the bad. Isn't that just an endless loop of negative and positive outcomes, in which humanity is eventually frayed beyond repair? Kelly disagrees:

> I don't think technology is neutral or a wash of good and bad effects. To be sure it does produce both problems and solutions, but the chief effect of technology is that it produces more possibilities. More options. More freedom, essentially. That's really good. That is the reason why people move to cities—for more choices.[3]

So, those of us who think the answer to the technological onslaught is to slow things down might want to think again. In *What Technology Wants*, Kelly makes quick work of both the Unabomber and the Amish, whose resistance to the growth of technology is futile, or even illusory. The Unabomber depended on bombs and the US mail system to attack technology; the Amish depend on hand tools that are, in turn, produced in high-tech factories.

Where it gets discomfiting, however, is when Kelly insists on technology's all-consuming nature. "It is an ever-elaborate tool that we wield and continually update to improve our world; and it is an ever-ripening superorganism, of which we are but a part, that is following a direction beyond our own making. Humans are both master and slave to the technium [his word for the technological universe], and our fate is to remain in this uncomfortable dual role."[4]

There is no way back, only through. Kelly admonishes us to "align ourselves with the imperative of the technium" because to do otherwise would be to "resist our second self."[5] Humanity and technology—like humanity and the zombies—are ultimately indistinguishable. "The conflict that the technium triggers in our hearts is due to our refusal to accept our nature—the truth is that we are continuous with the machines we create. . . . When we reject technology as a whole, it is a brand of self-hatred."[6]

But isn't the acceptance of humanity as a component part of technology also a form of self-hatred? Kelly sees a single thread of self-generation tying together the cosmos, the bios, and the technos into one act of creation. "Humans are not the culmination of this trajectory but an intermediary, smack in the middle between the born and the made."[7] We must either accept technology as our inevitable offspring and successor, or "reject technology as a whole." In Kelly's schema, there is no sustainable happy medium. Isn't there the possibility of a less dramatic, less apocalyptic middle ground?

In the apocalyptic scenario, we are either to hope for benevolence when our creation overtakes us or to negotiate with technology now in order to get some of what *we* want along with what *it* wants. As I have come to understand technology, however, it wants only whatever we program into it. I am much less concerned with whatever it is technology may be doing to people than what people are choosing to do to one another *through* technology. Facebook's reduction of people to predictively modeled profiles and investment banking's convolution of the marketplace into an algorithmic battleground were not the choices of machines but of humans.

Those who choose to see technology as equal to life end up adopting a "let it rip" approach to its development that ignores the biases of the many systems with which technology has become intertwined. The answer to the problems of technology is always just more technology, a pedal-to-the-metal ethos that is entirely consonant with laissez-faire capitalism. Ever since the invention of central currency, remember, the requirement of capitalism is to grow. It should not surprise us that in a capitalist society we would conclude that technology also *wants* to grow and that this growth supports the universe in its inexorable climb toward greater states of complexity.

However, I find myself unable to let go of the sense that human beings are somehow special, and that moment-to-moment human experience contains a certain unquantifiable essence. I still suspect there is something too quirky, too paradoxical, or too interpersonal to be imitated or re-created by machine life. Indeed, in spite of widespread confidence that we will crack the human code and replicate cognition within just a couple of decades, biology has a way of foiling even its most committed pursuers. The more we learn about DNA and the closer we come to mapping the entire genome, for example, the more we learn how small a part of the total picture it composes. We are no more determined by the neatly identifiable codons of the double helix than we are by the confused protein soup in which it actually operates. Put the same codons in a different person or species, and you'll get very different results. Our picture of human cognition is even hazier, with current psychopharmacology taking a shotgun approach to regulating neurotransmitters whose actual functioning we have only begun to understand. At our current level of technological sophistication, to argue that a virtual Second Life* simulation will soon become indistinguishable from real life smacks of fantasy and hubris.

Yet we are supposed to believe. Resistance to the logic and inevitability of the singularity is cast as quasi-religious. Nonbelievers are thought to be succumbing to a romantic notion of humanity

---

* An online virtual world

that is both steeped in morality and also in conflict with the scientific atheism we should have accepted by now. "What lies at the heart of every living thing is not a fire, not warm breath, not a 'spark of life,'" evolutionary biologist and outspoken atheist Richard Dawkins wrote in 1986, "it is information, words, instructions."[8] Evolution itself is to be viewed as an exchange of information between organism and environment. Or as science writer James Gleick chronicles in his book *The Information*, the universe itself is merely information reaching toward greater states of complexity. Atoms, matter, life, and technology are all just media for this information to evolve.

It seems to me this perspective has the medium and the message reversed. We humans are not the medium for information; information is a medium for humans. We are the content—the message.

It's an easy mistake to make, especially when we no longer have secure grounding in our past or a stable relationship to the future. Present shock is temporally destabilizing. It leads us to devalue the unbounded, ill-defined time of *kairos* for the neat, informational packets of *chronos*. We think of time as the numbers on the clock, rather than the moments they are meant to represent. We have nothing to reassure ourselves. Without a compelling story to justify a sustainable steady state for our circumstances, we jump to conclusions—quite literally—and begin scenario planning for the endgame.

In light of our rapid technological progress, this is not an altogether unexpected response. Thanks to self-replicating technologies such as computers, nanomachines, robots, and genomics, the future does seem to be upon us. It feels as if we can see the writing on the wall as it rapidly approaches from the distance. What is too easy to forget is that *we* are the ones simultaneously scrawling that very writing. We are the ones now writing the programs that will execute at some point in the future. We are the ones embedding our future reality with the values we want reaching back to us from there.[9] Truly living in this present becomes a form of time travel, in which everything we do actually matters to both our memory of the liquid past and, more important, the character of the unformed future.

Apocalypto relieves us of this responsibility by granting not only aspiration to technology but also superiority. We are freed of morality—that uniquely human trait—as well as the way it hampers our decision making and limits our choices. Technology will just go ahead and do it for us or, better, to us. Get it *over* with already. Like parents consoling themselves about their own imminent demise, we look to technology not merely as our replacement but as our heir.

But the reports of our death may be greatly exaggerated.

EVERYTHING OLD IS NEW AGAIN

Like followers of a new religion, devotees of the singularity have come to believe that their version of the apocalypse bears no resemblance to the many that have come before, that it is scientifically justified, and that the only choice that's left for us to make is to barrel ahead.

They may be right.

But this urgency to envision an imminent endgame is more characteristic of the religious tradition than the scientific one. And the extent to which we believe the harbingers of doom and rebirth has generally depended on the extent to which we feel dislocated from meaning and context. When persecution, torture, and killing in the Jewish ghettos reached their peak in medieval Europe, the Jews alleviated their despair with a form of messianism called the Kabbalah. In seventeenth-century England, the Puritans, their faith challenged by a political conflict between the king and the Roman Church, arose to realize the Revelations and bring about the "true kingdom of Christ." They colonized America with the express intent of bringing on the eschaton. In the 1820s, the terror and uncertainty of life on the frontier led many Americans to join a Christian revival movement called the Second Great Awakening, which included the Mormons, the Baptists, and the Shakers, and meant to prepare people for the imminent Second Coming. In just one of many examples, over one hundred thousand members of the Millerite movement prepared themselves for Christ's return—scheduled to happen by March 21, 1844—by selling their earthly goods and waiting in their attics. On March 22 the movement became known as the Great Disappointment and, later, the Seventh-Day Adventists.

Present shock provides the perfect cultural and emotional pretexts for apocalyptic thinking. It is destabilizing; it deconstructs the narratives we use to make meaning; it leads us to compulsively overwind, magnifying the stakes of any given moment; it leads us to draw paranoid connections where there are none; and, finally, its lack of regard for beginnings and endings—its focus on the perpetual now—drives us to impose order on chaos. We invent origins and endpoints as a way of bounding our experience and limiting the sense of limbo. We end up with no space—no time—between the concept of choosing to wear contact lenses and that of replacing our brains with nanobots. There is no continuum. Every tiny alpha must imply a terminal omega.

So which is more probable: That today's atheist apocalyptans are unique and right? Or that they are like their many predecessors—at the very least, in their motivations? If anything, the vehemence with which the believers in emergent complexity debunk all religion may betray their own creeping awareness of the religious underpinnings and precedents for their declarations.

In fact, the concept of Armageddon first emerged in response to the invention of monotheism by the ancient Persian priest Zoroaster, around the tenth or eleventh century BCE. Until that time, the dominant religions maintained a pantheon of gods reigning in a cyclical precession along with the heavens, so there was little need for absolutes. As religions began focusing on a single god, things got a bit trickier. For if there is only one god, and that god has absolute power, then why do bad things happen? Why does evil still exist?

If one's god is fighting for control of the universe against the gods of other people, then there's no problem. Just as in polytheism, the great achievements of one god can be undermined by the destructive acts of another. But what if a religion, such as Judaism of the First and Second Temple era, calls for one god and one god *alone*? How do its priests and followers explain the persistence of evil and suffering?

They do it the same way Zoroaster did: by introducing *time* into the equation. The imperfection of the universe is a product of its incompleteness. There's only one true god, but he's not done yet. In the monotheist version, the precession of the gods was no longer a continuous cycle of seasonal deities or metaphors. It was now a linear story with a clear endpoint in the victory of the one true and literal god. Once this happens, time can end.[10]

Creation is the Alpha, and the Return is the Omega. It's all good.

This worked well enough to assuage the anxieties of both the civilization of the calendar and that of the clock. But what about us? Without time, without a future, how do we contend with the lingering imperfections in our reality? As members of a monotheist culture—however reluctant—we can't help but seek to apply its foundational framework to our current dilemma. The less aware we are of this process—or the more we refuse to admit its legacy in our construction of new models—the more vulnerable we become to its excesses. Repression and extremism are two sides of the same coin.

In spite of their determination to avoid such constructs, even the most scientifically minded futurists apply the Alpha-Omega framework of messianic time to their upgraded apocalypse narratives. Emergence takes the place of the hand of God, mysteriously transforming a chaotic system into a self-organized one, with coherence and cooperation. Nobody seems able to explain how this actually happens.

Materialists argue that it's all just code. Each of the cells in an organism or the termites in a colony is following a set of very simple rules. When enough members are following that rule set, a bigger phenomenon becomes apparent—like the kaleidoscopic patterns in a Busby Berkeley dance routine, the coordination of a bee hive, or even the patterns in a sand dune created by regular ocean waves. That's why emergence enthusiasts believe computers should be able to take a set of simple rules and reconstruct the universe, humans and all.

But the moment when a system shifts from one level of organization to another—the moment when emergent behavior develops—is still very poorly understood, and almost as slippery a concept as biblical creation. This is not to say emergence is not a real and observable phenomenon, but only that we are still grappling with just when, why, and how it happens, as well as how much of a role our own subjectivity plays in judging a system to be complex or random. For humans, order usually means something that looks like ourselves.

So, the new myth goes—driven by technology and information's *need* to find ever-more complex expressions—our microchips get faster and more interconnected until they emerge as an independently complex system, one level higher than our own. All they needed were their simple, binary commands and a whole lot of feedback and iteration. They may or may not re-create humanity as a running virtual program, but they will have the codes to do so if they choose. If we are good, or prepare properly, we get a Second Life. In any case, what we think of as biological time will be over.

The singularity realizes the second coming of the Big Bang—the Omega point that completes human existence and ends the illusion of linear time. It may be upon us, or, as the illusory nature of time suggests, it may have already happened and we're just catching up with it. Imminent or not, it shares more characteristics with religion than its advocates like to admit, complete with an Omega point, a second life, an act of creation, a new calendar, and the dogged determination to represent itself as a total departure from all that came before.

This response is a bit more like that of the impatient, reactive Tea Partier than that of the consensus-building Occupier. More like the "inbox zero" compulsive than the person who answers email if and when he feels like it. More the hedge fund trader looking to see how many algorithms can dance on the head of a temporal pin than the investor looking for a business to capitalize over time. More the fractalnoid conspiracy theorist than the pattern recognizer.

The more appropriate approach to the pressures of apocalypto may be to let up on the pedal just a bit. This doesn't mean stopping altogether or stepping on the brakes. It could mean making sure we understand the difference between a marketplace that has been designed to accelerate no matter what and a reality that may or may not share this embedded agenda. It could mean beginning to envision slow paths to sustainability that don't require zombies or the demise of a majority of the world's population.

Most of all, as when confronting any of the many faces of present shock, it means accepting responsibility and dominion over the moment in which we are living right now. This has proved the hardest part for me and, I imagine, for you. I understand that just reading this book required you to carve out a big chunk of time, and to withhold this time from the many people, tasks, and Tweets vying for your immediate attention. I imagine it took more effort than reading a book of this length and depth would have required, say, ten years ago.

Likewise, although this is hardly the longest book I have written, it has been by far the hardest to complete. And I don't think that's because the subject was intrinsically more difficult or complex (or that I have grown more feeble), but because the environment in which I'm writing and my audience is living has changed. It is not you or I or the information that's so different, but the media and culture around us all.

At one point, I began developing marginalia alongside the text through which I could chronicle whatever it was I was *supposed* to be doing instead of writing at that moment. I maintained a separate vertical column on the edge of the page filled with the lunches I'd turned down; the emails that went unanswered; the missed offers to earn a buck; and all the interviews, articles, and appearances that could have led to something. But that list soon took up more space than the main body of text, so I stopped before it demoralized me into paralysis.

As I continued on, head down, I began to think more of the culture to which I was attempting to contribute through this work. A book? Really? How anachronistic! Most of my audience—the ones who agree with the sentiments I am expressing here—will not be getting this far into the text, I assure you. They will be reading excerpts on BoingBoing.net, interviews on Shareable.net, or—if I'm lucky—the review in the *New York Times*. They will get the gist of the argument and move on.

Meanwhile, in the years it has taken me to write this book—and the year after that to get it through the publishing process—I could have written dozens of articles, hundreds of blog posts, and thousands of Tweets, reaching more people about more things in less time and with less effort. Here I am writing opera when the people are listening to singles.

The solution, of course, is balance. Finding the sweet spot between storage and flow, dipping into different media and activities depending on the circumstances. I don't think I could have expressed present shock in a Tweet or a blog post or an article, or I would have. And taking the time to write or read a whole book on the phenomenon does draw a line in the sand. It means we can stop the onslaught of demands on our attention; we can create a safe space for uninterrupted contemplation; we can give each moment the value it deserves and no more; we can tolerate uncertainty and resist the temptation to draw connections and conclusions before we are ready; and we can slow or even ignore the seemingly inexorable pull from the strange attractor at the end of human history. For just as we can pause, we can also *un*-pause.

Thanks for your time.

ACKNOWLEDGMENTS

Acknowledgments are usually remembered more by those left out than those included. My accidental omissions will no doubt be compounded here, as this book has been in the works one way or another since around 1993. So to everyone who helped in one way or another over the past twenty years, thanks. You know who you are and I hope you see your impact on this work.

Were it not for my current agents, Katinka Matson, John Brockman, and Max Brockman, I may not have had the courage to put forth this sort of book in the first place. Your faith in my ability to dig deeper and write the books I was born to write has made me a better, smarter author. And thanks to Russell Weinberger, for bringing my books to the world.

As for you, Niki Papadopoulos, we may have been brought together by circumstance, but sometimes the Fates really do make their will known. It's great to have a partner in crime, especially in a world where new ideas are all but verboten. Thanks also to Gillian Blake, who originally pushed me to pursue presentism as a book, David Moldawer for making a home for it, and Courtney Young for encouraging me to keep considering the positive implications of present shock.

Thanks also to Natalie Horbachevsky, whose care and diligence made the editorial process painless, and to my extremely promising publicists Christine D'Agostini and Whitney Peeling, whom I am thanking in advance for having made sure you know this book exists. If early indications are to be trusted, we are all in good hands.

I had research help along the way as well, most notably from Rachel Rosenfelt, now editor of *TheNewInquiry*, and my boss of sorts. She found books and articles I wouldn't have on my own, and she acted so enthusiastic about them that I actually read them without being forced. Andrew Nealon also helped me organize my thoughts and references during the early stages of preparing the manuscript. He did so without judgment, which was both necessary and supportive during that floundering period.

Frank Kessler, Joost Raessens, and Mirko Schäfer, of Utrecht University, waited patiently for various drafts of my doctoral dissertation while I wrote this book (and, to be honest, the six books preceding this one). Working with them these past ten years upped my game considerably and brought much greater rigor to my writing. Thanks also to media ecologists everywhere, who helped me to understand how media create environments and why this matters.

Ian and Britta Alexander of EatMedia provided the physical and, in some cases, emotional space I needed to write this book. In addition to providing me an office, a computer, and connectivity, Ian fended off distractions as they came through the door and over the Internet. You are true friends. Brian Hughes, the latest member of the EatMedia family, proofread my text and got me through the final stages of manuscript production. He smiled the entire time.

Zach Sims and Ryan Bubinski, founders of Codecademy, deserve special thanks. They extended their invitation for me to work with them bringing code literacy to the world, and then waited patiently for me to finish this book before I could begin. They understand better than almost anyone how the best way to navigate the increasingly programmed landscape ahead is to know something about programming, oneself. Having their adventure ahead served as a light at the end of the long tunnel.

Many people have engaged with me about the ideas in this book. These ideas are as much yours as they are mine. You are, in an order that makes sense to me on a fractalnoid level, Dr. Mark Filippi, Ryan Freilino, Jerry Michalski, Kevin Slavin, Curtis Faith, Howard Rheingold, Terence McKenna, Stewart Brand, Ken Goldberg, Clay Shirky, Amber Case, Cintra Wilson, Jonathan Lethem, Samantha Hinds, David Bennahum, Walter Kirn, Steven Bender, Jeff Newelt, Barak Goodman, Rachel Dretzin, David Pescovitz, Janet Sternberg, Lance Strate, Mark Stahlman, Paul Levinson, Alan Burdick, Renee Hobbs, Nathalis Wamba, and Hermenauts everywhere. Thanks also to my mother, Sheila, who passed away before I started this one but always thought it was "a good idea."

Finally, and most important, my ever-supportive wife, Barbara. You have lived through enough of these now to know I will make it through—even when I don't. Thanks for knowing this would work, and agreeing to live with a crazy author in the first place. Love, Douglas.

NOTES

//

PREFACE

1. The number of births reported in New York hospitals nine months after 9/11 increased by 20 percent. For all the statistics, see Tom Templeton and Tom Lumley, "9/11 in Numbers," *Observer*, August 17, 2002.

2. As *Huffington Post* writer Lee Harrington explained, "I always told myself that 'one day' my marriage would get better. Then one day the future stopped." See Lee Harrington, "Falling Man Helped Me Face My Own Fears," *Huffington Post*, September 7, 2011, www.huffingtonpost.com/lee-harrington/falling-man-marriage_b_951381.html.

3. See any of the dozens of websites and publications that have popped up to debunk climate change, such as GlobalWarmingHysteria.com or Global WarmingLies.com, or S. Fred Singer, *Hot Talk Cold Science: Global Warming's Unfinished Debate* (Oakland, CA: The Independent Institute, 2001). For an account of the hacked global warming emails, begin with Katherine Goldstein and Craig Kanalley, "Global Warming Emails: Hack Raises Ethical Questions, Hoax and Scam Claims," *Huffington Post*, March 18, 2010, www.huffingtonpost .com/2009/11/23/global-warming-emails-hac_n_367979.html.

//

CHAPTER 1: NARRATIVE COLLAPSE

1. Mark Turner, *The Literary Mind* (New York: Oxford University Press, 1998).

2. Author Ursula K. Le Guin, 1979, quoted on www.qotd.org.

3. Alvin Toffler, "The Future as a Way of Life," *Horizon* 7 (3), 1965.

4. Ibid.

5. Joseph Campbell, *The Hero with a Thousand Faces* (Princeton, NJ: Princeton University Press, 1968).

6. Aristotle, quoted in Robert McKee, *Story: Substance, Structure, Style, and the Principles of Screenwriting* (New York: ReganBooks, 1997).

7. See my book *Playing the Future* (New York: HarperCollins, 1996) for more on the narrative novelty of these sorts of TV shows and movies, though with different conclusions about their significance.

8. Hampton Stevens, "The Meta, Innovative Genius of 'Community,'" *Atlantic*, May 12, 2011.

9. Zadie Smith quoted by James Wood in a review of *White Teeth*, "Human, All Too Inhuman: The Smallness of the 'Big' Novel," New Republic Online, July 24, 2000, www.tnr.com/article/books-and-arts/human-all-too-inhuman.

10. Mike Freeman, "Saints Took Common Practice of Bounties to a New, Dangerous Level," CBSsports.com, March 5, 2012.

11. All three quotes are from Jon Baskin, "Steroids, Baseball, America," *Point* 4 (Spring 2011).

12. Bill Simmons, "Confronting My Worst Nightmare," ESPN.com, May 9, 2009.

13. Murray Chass, "On Baseball; Senate Posse Is Passing Steroid Buck to Baseball," *New York Times*, March 16, 2004.

14. Tom Van Riper, "MLB Faces Fourth Straight Attendance Decline," *Forbes*, September 7, 2011; Associated Press, "NFL Ticket Sales Decline for Third Straight Year," September 8, 2010.

15. Terri Judd, "Teenagers Risk Death in Internet Strangling Craze," *Independent*, January 6, 2010.

16. Walter Lippmann, *Public Opinion* (New York: Free Press, 1922).

17. Steven Livingston, "Clarifying the CNN Effect: An Examination of Media Effects According to Type of Military Intervention," PDF, John F. Kennedy School of Government's Joan Shorenstein Center on the Press, Politics and Public Policy at Harvard University, 1997.

18. For a look at Luntz's process, see my PBS *Frontline* documentary, "The Persuaders" (2004).

19. "Transcript of President Bush's Prayer Service Remarks," National Day of Prayer and Remembrance for the Victims of the Terrorist Attacks on September 11, 2001, Washington National Cathedral, September 14, 2001, www.opm.gov/guidance/09-14-01gwb.htm.

20. From 1985 to 2005, the number of Americans unsure about evolution increased from 7 percent to 21 percent, according to a study by the National Center for Science Education, cited in "Public Praises Science; Scientists Fault Public, Media," Pew Research Center, July 9, 2009, www.people-press.org.

21. According to Gallup data as of 2010, 48 percent of Americans believe the seriousness of global warming is exaggerated, up from 31 percent in 1997. Frank Newport, "Americans' Global Warming Concerns Continue to Drop," *Gallup Politics*, March 11, 2010, www.gallup.com.

22. Richard Edelman in Sheldon Rampton and John Stauber, *Trust Us, We're Experts: How Industry Manipulates Science and Gambles with Your Future* (New York: Tarcher/Penguin, 2002).

23. Lymari Morales, "In U.S., Confidence in Newspapers, TV News Remains a Rarity," *Gallup Politics*, August 13, 2010, www.gallup.com/poll/142133/confidence-newspapers-news-remains-rarity.aspx.

24. Pew Research Center for the People & the Press, September 22, 2011, www.people-press.org.

25. Kasun Ubayasiri, "Internet and the Public Sphere: A Glimpse of YouTube," Central Queensland University, 2006, and updated, on EJournalist.com, http://ejournalist.com.au.

26. Andrew Keen, *The Cult of the Amateur* (New York: Crown, 2007), 48.

27. Mark Lilla, "The Tea Party Jacobins," *New York Review of Books*, May 27, 2010.

28. David Frum, "When Did the GOP Lose Touch with Reality?" *New York*, November 20, 2011.

29. Tommy Christopher, "Van Susteren Explains Why Anti-Fox Clip with Occupy Wall St. Protester Got Cut," MediaIte.com, October 3, 2011, www.mediaite.com/tv/van-susteren-explains-why-anti-fox-interview-with-occupy-wall-st-protester-got-cut.

///

CHAPTER 2: DIGIPHRENIA: BREAKING UP IS HARD TO DO

1. See Nicholas Carr, *The Shallows: What the Internet Is Doing to Our Brains* (New York: W. W. Norton, 2010) and Sherry Turkle, *Alone Together: Why We Expect More from Technology and Less from Each Other* (New York: Basic Books, 2011).

2. Look at Aristotle and Thomas Aquinas for the earliest notions of real time and existence compared with human-defined days and years. See Lewis Mumford, Harold Innis, David S. Landes, and Jeremy Rifkin for understandings that bring us through the Industrial Age and even into digital culture. My own historical and theoretical frameworks build on their works.

3. For more on this pre–Axial Age spiritual outlook, see Karen Armstrong's *A History of God* (New York: Ballantine Books, 1993).

4. Walter Ong, *Orality and Literacy* (New York: Routledge, 2002), 69.

5. See my book *Nothing Sacred* (New York: Crown, 2003).

6. Jeremy Rifkin, *Time Wars: The Primary Conflict in Human History* (New York: Touchstone Books, 1989).

7. David S. Landes, *A Revolution in Time: Clocks and the Making of the Modern World* (Cambridge: Harvard University Press, 1983), 10.

8. Robert Levine, *A Geography of Time: The Temporal Misadventures of a Social Psychologist, or How Every Culture Keeps Time Just a Little Bit Differently* (New York: Basic Books), 1997.

9. David Montgomery, *The Fall of the House of Labor: The Workplace, the State, and American Labor Activism, 1865–1925* (Cambridge: Cambridge University Press, 1989).

10. Mark P. McDonald, PhD, "The Nature of Change Is Changing: The New Pattern," April 12, 2010, http://blogs.gartner.com/mark_mcdonald/2010/04/12/the-nature-of-change-is-changing-the-new-pattern/.

11. Dave Gray, "Change Is Changing," October 26, 2011, www.dachisgroup.com/2011/10/change-is-changing/.

12. Stefanie Luthman, Thomas Bliesener, Frithjof Staude-Müller, "The Effect of Computer Gaming on Subsequent Time Perception," *Journal of Psychosocial Research in Cyberspace* 3 (1), June 2009.

13. Rebecca Maksel, "When Did the Term 'Jet Lag' Come into Use?" *Air & Space Magazine*, June 18, 2008. Interestingly, it was subsequently discovered that west-to-east travel is much more challenging to biorhythms than east-to-west. The inaccuracy of the initial study's results may be because they used only four test subjects.

14. Ibid.

15. SAIC Information Services, "NASA and the FAA: A Technology Partnership for the New Millennium," www.aeronautics.nasa.gov/docs/chicago/fcp.htm.

16. This is a new but scientifically researched field. See Roger Lewin, *Making Waves* (Emmaus, PA: Rodale, 2005).

17. Maggie Fox, "Shift Work May Cause Cancer, World Agency Says," Reuters, November 30, 2007.

18. Clay Shirky, *Cognitive Surplus: Creativity and Generosity in a Connected Age* (New York: Penguin, 2010).

19. Lisa Napoli, "As If in a Seller's Dream, the Bags Fly Out of the Studio," *New York Times*, December 7, 2004.

20. Ibid.

21. Gary Wolf, "Tim Ferriss Wants to Hack Your Body," *Wired*, December 2010.

22. By 2007, 43 percent of email users said the first thing they do when they wake is check for new messages. AOL study cited in "Email Statistics," at http://powerprodirect.com.

23. James Bridle, "The New Aesthetic: Waving at the Machines," talk delivered at Web Directions South, Sydney, Australia, December 5, 2011, http://booktwo.org/notebook/waving-at-machines.

24. Vanessa Grigoriadis, "Everybody Sucks: Gawker and the Rage of the Creative Underclass," *New York*, October 14, 2007.

25. For Dardik's story and techniques, see Roger Lewin, *Making Waves: Irving Dardik and His Superwave Principle* (Emmaus, PA: Rodale, 2005).

26. Most of this work is unpublished, but for more on his approach, see David Alan Goodman, "Declare Your Independence," *Scientist* 17 (12), June 16, 2003, p. 13.

27. Joel C. Robertson, *Natural Prozac: Learning to Release Your Body's Own Anti-Depressants* (New York: HarperOne, 1998).

28. All quotes from Mark Filippi are from interviews I conducted with him in February and March 2012.

29. "Tensegrity" is a term used most famously by Buckminster Fuller to describe the structural integrity of various systems. See R. Buckminster Fuller, "Tensegrity," 1961, at www.rwgrayprojects.com/rbfnotes/fpapers/tensegrity/tenseg01.html.

30. See http://Lifewaves.com or http://somaspace.org.

31. Steven Johnson, *Where Good Ideas Come From* (New York: Riverhead, 2010).

32. Kutcher Tweeted in defense of a fired college football coach, only learning later that the coach had covered up a child molestation. Kutcher then turned his Twitter account over to a professional PR firm.

33. For more on drones and drone pilots, see my PBS documentary on *Frontline*, *Digital Nation*, in particular the section on "War by Remote," www.pbs.org/wgbh/pages/frontline/digitalnation/blog/2009/10/new-video-fighting-from-afar.html. Also see Phil Stewart, "Overstretched Drone Pilots Face Stress Risk." Reuters, December 18, 2011.

34. See Nicholas Carr, *The Shallows*; Sherry Turkle, *Alone Together*; and Maggie Jackson, *Distracted: The Erosion of Attention and the Coming Dark Age* (New York: Prometheus, 2009).

35. Henry Greely, "Towards Responsible Use of Cognitive-enhancing Drugs by the Healthy," *Nature*, December 7, 2008.

36. James G. March, *A Primer on Decision Making* (New York: Free Press, 1994), 245.

37. James Borg, *Body Language: 7 Easy Lessons to Master the Silent Language* (Upper Saddle River, NJ: FT Press, 2010).

38. Robert McKee and Dr. David Ross, "From Lean Manufacturing to Lean Supply Chain: A Foundation for Change," a whitepaper for Lawson, available at http://swe.lawson.com/www/resource.nsf/pub/Lawson_Whitepaper_2_A4_LowRes.pdf/$FILE/Lawson_Whitepaper_2_A4_LowRes.pdf.

CHAPTER 3: OVERWINDING: THE SHORT FOREVER

1. David Hess, "The NBA Lockout Has Increased Injury Rates," *Notes from the Sports Nerds* blog, February 7, 2012, www.teamrankings.com/blog/nba/the-nba-lockout-has-increased-injury-rates.

2. Freeman Dyson, *From Eros to Gaia* (New York: Pantheon, 1992), 341.

3. Stewart Brand, *The Clock of the Long Now: Time and Responsibility* (New York: Basic Books, 1999), 49.

4. Alfred Korzybski, *Science and Sanity: An Introduction to Non-Aristotelian Systems and General Semantics* (Lakeville, CT: International Non-Aristotelian Library Pub.; distributed by the Institute of General Semantics, 1958), 376.

5. Ibid.

6. Productivity guru Merlin Mann is the originator of the term "inbox zero," but the idea was first posed by Mark Hurst in his 2007 book *Bit Literacy*. I have used both their systems with success but have found myself differing with them on the question of email—most likely because email and the ways we receive it have changed over the past decade.

7. Mark Hurst, *Bit Literacy: Productivity in the Age of Information and E-mail Overload* (New York: Good Experience, 2007).

8. For more on this, see my book *Life Inc.* (New York: Random House, 2009).

9. See Bernard Lietaer, "Complementary Currencies in Japan Today," *International Journal of Community Currency Research* 8 (2004).

10. The Renfrew Center Foundation for Eating Disorders, "Eating Disorders 101 Guide: A Summary of Issues, Statistics and Resources," published September 2002, revised October 2003, www.renfrew.org.

11. Ibid.

12. Adam Sternbergh, "Up with Grups," *New York*, March 26, 2006.

13. Ibid.

14. Search YouTube for "Smells Like a Feeling."

15. Zachary Lazar, "The 373-Hit Wonder," *New York Times*, January 6, 2011.

16. Jeffrey Rosen, "The Web Means the End of Forgetting," *New York Times*, July 21, 2010.

17. Opera Solutions website, www.operasolutions.com/about-us.

18. Stephanie Clifford, "Thanksgiving as Day to Shop Meets Rejection," *New York Times*, November 10, 2011.

19. Bill Gentner, senior vice president for marketing, quoted in Clifford, ibid.

20. Richard Barbrook, *Imaginary Futures: From Thinking Machines to the Global Village* (London: Pluto, 2007).

21. John Hagel, "The 2011 Shift Index: Measuring the Forces of Long-Term Change," *Deloitte & Touche—Edge Report, 2011*, www.deloitte.com/us/shiftindex.

22. See my book *Life Inc.* (New York: Random House, 2009), 120.

23. Liz Moyer, "Fund Uses Behavioral Finance to Find Value Plays," CBS MarketWatch, June 28, 2011, www.marketwatch.com.

24. Uttara Choudhury, "Behavioral Economics has Never Been Hotter," Brain gainmag.com.

25. Robert D. Manning, *Credit Card Nation: The Consequences of America's Addiction to Credit* (New York: Basic Books, 2000).

26. "Corelogic Reports Negative Equity Increase in Q4 2011," *BizJournals*, March 1, 2012, http://assets.bizjournals.com/orlando/pdf/CoreLogic%20underwater%20mortgage%20list.pdf. Also: "Despite Home Value Gains, Underwater Homeowners Owe $1.2 Trillion More than Homes' Worth," *Zillow Real Estate Research*, May 24, 2012, www.zillow.com/blog/research/2012/05/24/despite-home-value-gains-underwater-homeowners-owe-1-2-trillion-more-than-homes-worth.

27. For a quick explanation and confirmation of these facts, see Serena Ng and Carrick Mollenkamp, "Goldman Fueled AIG Gambles," *Wall Street Journal*,

December 12, 2009. For a lengthy but compelling account of the entire Goldman Sachs saga, see Gretchen Morgenson and Joshua Rosner's *Reckless Endangerment: How Outsized Ambition, Greed, and Corruption Led to Economic Armageddon* (New York: Times Books, 2011).

28. See gaming and Internet analyst Kevin Slavin's terrific presentation on this history to the Lift11 Conference at www.livestream.com/liftconference/video?clipId=pla_08a3016b-47e9-4e4f-8ef7-ce71c168a5a8.

29. Kevin Slavin, "How Algorithms Shape Our World," TedTalks, July 2011, www.ted.com/talks/kevin_slavin_how_algorithms_shape_our_world.html.

30. Nina Mehta, "Automatic Futures Trade Drove May Stock Crash, Report Says," *Bloomberg Businessweek*, October 4, 2010. See also Graham Bowley. "Lone $4.1 Billion Sale Led to 'Flash Crash' in May," *New York Times*, October 1, 2010.

31. Brian Bremner, "The Bats Affair: When Machines Humiliate their Masters," *Bloomberg Businessweek*, March 23, 1012, www.businessweek.com/articles/2012-03-23/the-bats-affair-when-machines-humiliate-their-masters.

32. For the basics, see Alexandra Zendrian, "Don't Be Afraid of the Dark Pools," *Forbes*, May 18, 2009.

33. John Henley, "Greece on the Breadline: Cashless Currency Takes Off," *Guardian*, March 16, 2012.

34. Ibid.

35. Eric Westervelt, "Fiscal Localism on Rise in Germany," NPR, *All Things Considered*, July 15, 2010.

36. Judson Green's history and philosophies are taught at the Disney Institute in Orlando, Florida, which I attended as part of my research for this book. For more, see The Disney Institute and Theodore Kinni, *Be Our Guest: Perfecting the Art of Customer Service* (Glendale, CA: Disney Editions, 2011).

37. Michael McCarthy, "War of Words Erupts at Walt Disney," *USA Today*, December 2, 2003.

38. Dr. Ofer Merin, quoted in Catherine Porter, "Israeli Field Hospital Carries on Inspiring Work in Japan," *Toronto Star*, April 4, 2011.

39. Joichi Ito, "Innovating by the Seat of Our Pants," *New York Times*, December 6, 2011.

40. Robert Axelrod, *The Evolution of Cooperation* (New York: Basic Books, 1984).

CHAPTER 4: FRACTALNOIA: FINDING PATTERNS IN THE FEEDBACK

1. Steven Johnson, *Where Good Ideas Come From* (New York: Riverhead, 2010).

2. Kevin Dunbar, "How Scientists Build Models: InVivo Science as a Window on the Scientific Mind," www.utsc.utoronto.ca/~dunbarlab/pubpdfs/KDMBR99.pdf.

3. Kevin Roberts, interviewed in Barak Goodman, Rachel Dretzin, and Douglas Rushkoff, *The Persuaders*, PBS, *Frontline*, 2004.

4. "Chevy Tahoe, Trump Create Open Source Fun," *Oil Drum*, April 3, 2006, http://energyandourfuture.org/story/2006/4/3/164232/5126.

5. In a later riff on the same phenomenon, Shell's website for people to create advertisements promoting drilling for oil in the Arctic—http://arcticready .com—was eventually revealed to be a fake but not before hundreds of attack ads were created which utilities people thought had been provided by Shell. See the media-activist site http://YesLab.org for more on this.

6. See the website for the company at http://valvesoftware.com.

7. http://boingboing.net/2012/04/22/valve-employee-manual-describe.html.

8. Ibid.

9. László Mérő, *Moral Calculations* (New York: Springer-Verlag, 1998).

10. See my book *Life Inc.* (New York: Random House, 2009).

11. Archibald MacLeish, "Bubble of Blue Air," *New York Times*, December 25, 1968, p. 1.

12. Lenora Foerstal and Angela Gilliam, *Confronting Margaret Mead: Scholarship, Empire, and the South Pacific* (Philadelphia: Temple University Press, 1992), 126–27.

13. Steven Pinker, quoted in Nick Gillespie, "Hayek's Legacy," *Reason*, January 2005.

14. James Surowiecki, quoted in Gillespie, ibid.

15. See Manuel De Landa, *War in the Age of Intelligent Machines* (Cambridge, MA: MIT Press, 1992).

16. Jeff Sommer, "A Market Forecast That Says 'Take Cover,' " *New York Times*, July 3, 2010.

17. Philip E. Tetlock, *Expert Political Judgment: How Good Is It? How Can We Know?* (Princeton: Princeton University Press, 2006).

18. Walter Kirn, *Lost in the Meritocracy: The Undereducation of an Overachiever* (New York: Doubleday, 2009).

19. Richard Nisbett, quoted in Joshua Cooper Ramo, *The Age of the Unthinkable* (New York: Little, Brown, 2009).

20. Ramo, *Age of the Unthinkable*.

21. Ibid.

22. You can find out more or download the demo at www.thebrain.com.

23. You can see Jerry's Brain at http://jerrysbrain.com.

24. April Rinne and Jerry Michalski, "Polymaths, Bumblebees and the 'Expert' Myth," *Washington Post*, March 28, 2011.

25. Gordon Bell, Gordon Bell home page, http://research.microsoft.com/en-us/um/people/gbell/ (accessed August 11, 2011).

CHAPTER 5: APOCALYPTO

1. Mathew Barrett Gross and Mel Gilles, *The Last Myth* (Amherst, NY: Prometheus, 2012).

2. Rocco Castoro, "Ray Kurzweil: That Singularity Guy," *Vice*, April 1, 2009, www.vice.com.

3. John Brockman, "The Technium and the 7th Kingdom of Life: A Talk with Kevin Kelly," *Edge*, July 19, 2007, www.edge.org/3rd_culture/kelly07/kelly07_index.html.

4. Kevin Kelly, *What Technology Wants* (New York: Viking, 2010), 187.

5. Ibid., 188.

6. Ibid., 189.

7. Ibid., 356.

8. Richard Dawkins, *The Blind Watchmaker: Why the Evidence of Evolution Reveals a Universe without Design* (New York: W. W. Norton, 1986).

9. See my book *Program or Be Programmed* (New York: Or Books, 2010).

10. For a great chronicle and analysis of the apocalypse meme, see John Michael Greer, *Apocalypse Not* (Berkeley, CA: Cleis Press, 2011).

SELECTED BIBLIOGRAPHY

Axelrod, Robert. *The Evolution of Cooperation*. New York: Basic Books, 1984.

Barbrook, Richard. *Imaginary Futures: From Thinking Machines to the Global Village*. London: Pluto, 2007.

Boorstin, Daniel J. *The Image: Or, What Happened to the American Dream*. New York: Atheneum, 1962.

Brand, Stewart. *The Clock of the Long Now: Time and Responsibility*. New York: Basic Books, 1999.

Brockman, John. *Afterwords: Explorations of the Mystical Limits of Contemporary Reality*. New York: Anchor, 1973.

Carr, Nicholas G. *The Shallows: What the Internet Is Doing to Our Brains*. New York: W. W. Norton, 2010.

Carse, James P. *Finite and Infinite Games*. New York: Ballantine, 1986.

Dawkins, Richard. *The Selfish Gene*. Oxford: Oxford University Press, 1989.

De Landa, Manuel. *War in the Age of Intelligent Machines*. Cambridge, MA: MIT Press, 1992.

Eriksen, Thomas Hyllard. *Tyranny of the Moment: Fast and Slow Time in the Information Age*. London: Pluto Press, 2001.

Fogg, B. J. *Persuasive Technology: Using Computers to Change What We Think and Do*. San Francisco, CA: Morgan Kaufmann, 2003.

Fuller, R. Buckminster. *Operating Manual for Spaceship Earth*. Carbondale: Southern Illinois University Press, 1969.

Greer, John Michael. *Apocalypse Not: A History of the End of Time*. Berkeley, CA: Viva Editions, an Imprint of Cleis, 2011.

Gross, Mathew Barrett and Mel Gilles. *The Last Myth*. Amherst, NY: Prometheus, 2012.

Innis, Harold Adams. *Changing Concepts of Time.* Lanham, MD: Rowman & Littlefield, 2004.

Jacobs, Jane. *Systems of Survival: A Dialogue on the Moral Foundations of Commerce and Politics.* New York: Random House, 1992.

Johnson, Steven. *Where Good Ideas Come From.* New York: Riverhead, 2010.

Kelly, Kevin. *What Technology Wants.* New York: Viking, 2010.

Korzybski, Alfred. *Selections from Science and Sanity: An Introduction to Non-Aristotelian Systems and General Semantics.* 5th ed. Fort Worth, TX: Institute of General Semantics, 2005.

Kurzweil, Ray. *The Age of Spiritual Machines: When Computers Exceed Human Intelligence.* New York: Viking, 1999.

———. *The Singularity Is Near: When Humans Transcend Biology.* New York: Viking, 2005.

Landes, David S. *Revolution in Time: Clocks and the Making of the Modern World.* Cambridge, MA: Belknap of Harvard University Press, 1983.

Lanier, Jaron. *You Are Not a Gadget: A Manifesto.* New York: Alfred A. Knopf, 2010.

Lasch, Christopher. *The Culture of Narcissism: American Life in an Age of Diminishing Expectations.* New York: W. W. Norton, 1978.

Leach, William. *Land of Desire: Merchants, Power, and the Rise of a New American Culture.* New York: Pantheon, 1993.

Levine, Robert. *A Geography of Time: The Temporal Misadventures of a Social Psychologist, or How Every Culture Keeps Time Just a Little Bit Differently.* New York: Basic Books, 1997.

McLuhan, Marshall. *Understanding Media: The Extensions of Man.* Critical Edition. Corte Madera, CA: Gingko, 2003.

Mérő, László. *Moral Calculations: Game Theory, Logic, and Human Frailty.* New York: Copernicus, 1998.

Montgomery, David. *The Fall of the House of Labor: The Workplace, the State, and American Labor Activism, 1865–1925.* Cambridge: Cambridge University Press, 1989.

Mumford, Lewis. *Technics and Civilization.* Chicago: University of Chicago Press, 2010.

Postman, Neil. *Technopoly: The Surrender of Culture to Technology.* New York: Alfred A. Knopf, 1992.

Ramo, Joshua Cooper. *The Age of the Unthinkable: Why the New World Disorder Constantly Surprises Us and What to Do About It.* New York: Little, Brown and Company, 2009.

Rifkin, Jeremy. *Time Wars: The Primary Conflict in Human History.* New York: Henry Holt, 1987.

Rushkoff, Douglas. *Life Inc.: How the World Became a Corporation and How to Take It Back.* New York: Random House, 2009.

————. *Program or Be Programmed: Ten Commands for a Digital Age*. New York: Or Books, 2010.

Shirky, Clay. *Here Comes Everybody: The Power of Organizing Without Organizations*. New York: Penguin, 2008.

Strate, Lance. Introductory Note to "Eine Steine Nacht Muzak." *KronoScope* 10, no. 1–2 (2010).

Strate, Lance, Ronald L. Jacobson, and Stephanie B. Gibson. *Communication and Cyberspace: Social Interaction in an Electronic Environment*. 2nd ed. Cresskill, NJ: Hampton, 2003.

Turkle, Sherry. *Alone Together: Why We Expect More from Technology and Less from Each Other*. New York: Basic Books, 2011.

Wiener, Norbert. *The Human Use of Human Beings: Cybernetics and Society*. 2nd ed. Garden City, NY: Doubleday, 1954.

Zimbardo, Philip G., and John Boyd. *The Time Paradox: The New Psychology of Time That Will Change Your Life*. New York: Free Press, 2008.

INDEX

abstractions, 138–39
acting now, 159–69f
Acxiom, 158
advertising, 28, 128, 167, 210, 245.
    See also commercials; marketing/
    market research
age/aging, 149–52
agriculture, 100–101, 185
Agriculture Department, U.S., 170
air travel, 89–90
Al Jazeera, 216
algorithms, 8, 178–79, 180, 181,
    182, 183, 229, 257, 264
Alhurra (cable news channel), 216
Alpha-Omega framework, 262–64
"always-on," 1–2, 73–74, 85, 94–95,
    97–98, 186, 211
America: adolescence of, 45–46;
    futurism and character of, 12
American Dream, 12–13
An American Family (documentary), 35
American Idol (TV show), 37, 54, 213
American Psychiatric Association, 166
American Psychological
    Association, 38
animated TV shows, 23–24, 25–26
answering machines, 128
apocalypto: appropriate approach to,
    264–66; bunkers for, 243–45;

change and, 264–65; choices and,
256, 257–58, 260; conflation of
apocalypse scenarios and, 246;
definition/characteristics of,
2, 245, 261; digiphrenia and,
245; emergence of Armageddon
concept and, 261–62; fear of
the future and, 246; fractalnoia
and, 245–46, 264; human limits
and, 254–60; as manifestation
of present shock, 7, 243–66;
modern problems and, 246–47;
narrative collapse and, 245; new
"now" and, 3; overwinding and,
245, 261; transcending humanity
and, 251–54; zombies and,
247–50, 264
    See also specific topic
Apple Corporation, 13, 108, 111,
    167–68, 182, 203, 218, 239
Arab Spring, 52, 55, 203, 216
Aristotle, 19, 23
arms race, 178–79
art works, overwinding and, 153–55
artistic visionaries, 231–32
Asia, clocks and timing in, 80
athletes. See sports
attention, competition for, 124,
    265–66

Printed in the United States
by Baker & Taylor Publisher Services